Anti-Social Behaviour Orders

Anti-Social Behaviour Orders

A Culture of Control?

Jane Donoghue
University of Reading, UK

First published 2010 by
PALGRAVE MACMILLAN

Palgrave Macmillan in the UK is an imprint of Macmillan Publishers Limited, registered in England, company number 785998, of Houndmills, Basingstoke, Hampshire RG21 6XS.

Palgrave Macmillan in the US is a division of St Martin's Press LLC, 175 Fifth Avenue, New York, NY 10010.

Palgrave Macmillan is the global academic imprint of the above companies and has companies and representatives throughout the world.

Palgrave® and Macmillan® are registered trademarks in the United States, the United Kingdom, Europe and other countries.

ISBN 978-0-230-59444-9 hardback

This book is printed on paper suitable for recycling and made from fully managed and sustained forest sources. Logging, pulping and manufacturing processes are expected to conform to the environmental regulations of the country of origin.

A catalogue record for this book is available from the British Library.

A catalog record for this book is available from the Library of Congress.

10 9 8 7 6 5 4 3 2 1
19 18 17 16 15 14 13 12 11 10

Printed and bound in Great Britain by
CPI Antony Rowe, Chippenham and Eastbourne

Contents

v

Acknowledgements

Sincere thanks are due to all those who have so kindly given up their time to participate in my research. This includes solicitors, criminal justice and community safety practitioners, court staff and members of the judiciary. I am immensely grateful to all participants for being so generous with their time, for their patience and for their willingness to contribute to research in this area. I also wish to acknowledge the facilities provided by the University of Reading, and the hugely supportive and collegial environment at the School of Law, where I have been based while completing this manuscript. In particular, I wish to thank Professor Patricia Leopold, Professor Chris Hilson, Professor Sandy Ghandi and Dr Grace James. My thanks also to my friend and colleague, Professor Sandeep Gopalan, of the National University of Ireland, for his support and encouragement. Special thanks are also due to Professor Reece Walters, of The Open University, for the stimulating and inspiring discussions that we held during the early years of my research on anti-social behaviour. I am also very grateful to Philippa Grand and Olivia Middleton at Palgrave Macmillan for their patience and encouragement from the submission of my initial proposal through to the delivery of the final manuscript.

Finally, deepest gratitude is due to Moira and Brian Donoghue, for their endless support and kindness.

Preface

ASBOs: A Culture of Control? aims to provide a new contribution to the debate on anti-social behaviour policy in Britain. This book devotes itself to a *critical* reading of the existing scholarship on anti-social behaviour management and the use of ASBOs. It argues that much of the available literature on anti-social behaviour policy and the use of ASBOs has necessarily been impaired by too much ideological partisanship. As such, *ASBOs: A Culture of Control?* considers the representation of ASBOs within the academic literature and the inherent scepticism underpinning the production and reception of analyses on ASBOs within sociological and criminological writing.

Despite observations that anti-social behaviour was becoming increasingly politically sidelined, there have been recent new developments in anti-social behaviour policy. Chapter 2 discusses on the government's 'new drive' on anti-social behaviour and the proposed improvements to be made to anti-social behaviour management are outlined. In Chapter 3, the political background to the introduction of the ASBO is traced. The end of penal welfarism and the rise of the politics of law and order are discussed alongside arguments about the nature of anti-social behaviour policy as embodying principles associated with neo-liberalism and conservatism. In Chapter 4, readers are provided with an account of the historical landscape of anti-social behaviour. This chapter also consideres whether anti-social behaviour can be understood as a new 'moral panic' in contemporary Britain. Chapter 5 examines anti-social behaviour and the use of ASBOs specifically in relation to social housing. The pervasive academic preoccupation with embodying accounts of ASBOs within the domain of 'social control' theory is discussed. This chapter also considers whether anti-social behaviour interventions represent part of a broader historical continuum of moral censure in social housing. In Chapters 6 and 7, the legal and court processes impacting upon the use of ASBOs are then considered in detail. Specific aspects of legal procedure including the use of civil rules of evidence, the formation of prohibitions, breach proceedings, the use of interim orders and orders on conviction and the defence of ASBO action are examined. Drawing upon case law, relevant statutory provisions and empirical research, to what extent ASBOs are circumventing the principles and safeguards of due process will be considered. Finally, in Chapter 8, the book concludes that ASBOs should not be understood as part of a 'culture of control' but as underpinned by a process of social protection, of 'social aegis' which is principally concerned with increasing social capital, as opposed to the marginalisation and exclusion of 'targeted groups'.

1
Introduction

Although anti-social behaviour orders (ASBOs) are a relatively recent development in criminal justice policy in Britain, they have nonetheless been an extremely topical area of law and policy for some years. Despite limited empirical evidence on the quantifiable 'effectiveness' of the orders, research has consistently demonstrated a considerable widespread public support for the creation and use of ASBOs. In June 2005, for example, 82 per cent of respondents to a survey on ASBOs supported their use in tackling anti-social behaviour (Ipsos Mori, 2005). The reported level of importance attached by individuals and by communities to addressing and deterring anti-social behaviour swiftly and effectively (Home Office, 2006c; Ipsos Mori, 2005), and research evidence of both the economic and emotional cost to victims of anti-social behaviour (Upson, 2006) has meant that the government has repeatedly encouraged local authorities and the police to make greater use of the orders. However, the (continued) public support enjoyed by ASBOs is strongly antithetical to the dominant academic view held by many criminal justice elites who overwhelmingly characterise the use of ASBOs as indicative of a wider, increasingly authoritarian political and social agenda embodying marginalisation, social control and the 'management' of targeted groups and/or individuals.

There are relatively few academic books currently available specifically on the subject of anti-social behaviour policy and these are largely concerned with presenting a critical perspective on ASBOs. Similarly, journal articles and other academic literature are frequently characterised by a discussion of ASBOs (predominantly or exclusively) in pejorative terms of social control. For example, academic discussion of anti-social behaviour policy and the use of ASBOs inculcates such

1

terms as 'cosmopolitan intolerance' (Bannister et al., 2006), 'negative control' (Drakeford and McCarthy, 2000), 'precautionary injustice' (Squires and Stephen, 2005), 'symbolic punitiveness' (McIntosh, 2008), 'authoritarian agenda' (Brown, 2004), 'new punitiveness' (Pratt et al., 2005), 'reactive enforcement' (Squires, 2006), 'the criminalisation of public leisure' (Stephen, 2008), 'compulsory moral integration' (Gillies, 2005), 'marginalisation' (Brown, 2004) and 'exclusion' (Hester, 2000; Scraton, 2004). In this respect, academic literature on anti-social behaviour management repeatedly seeks to identify anti-social behaviour policy as – overwhelmingly – a means for domination, stratification and as an instrument for social exclusion. Indeed, Jacobson et al. (2008: 51) describe the New Labour administration's approach to anti-social behaviour management as supported by a 'depressing form of anti-intellectualism'.

Critical academic accounts of anti-social behaviour management and the use of ASBOs often cite Garland's thesis on the 'criminology of the other', in which the notion of crime as a normal/anticipated fact of everyday life is rejected. Instead, it is re-dramatised in 'melodramatic terms, viewing it as a catastrophe, framing it in the language of warfare and social defence' (Garland, 2001: 184). The acquiescent 'good citizen' is juxtaposed with the 'excluded and…embittered' criminological outsider (ibid: 137). In this way, ASBOs have been understood as forming part of the new 'culture of control' which Garland contends is distinguished by the (re)emergence of punitive sanctions, increased emphasis upon victims and the creation of policies aimed at greater community involvement in justice processes. According to Garland, following social and economic changes, crime control has become increasingly politicised, resulting in the dismantling of the 'penal welfare' approach so evident in criminal justice practice up until the 1970s, when the 'orthodoxies of rehabilitative faith collapsed in virtually all developed countries'. The replacement of penal welfarism by a 'culture of control' has thus resulted in bifurcated crime control policies which are divided between those adaptive strategies, focussing on community involvement in justice processes and those sovereign state strategies which are concerned with the domination and control of offenders. For many academic analyses, ASBOs are part of this new culture, which, as Zedner observes (2002: 356) 'focuses relentlessly on the policies and technologies of control'.

This book is distinct from others in the subject field and presents an alternative perspective on anti-social behaviour management which expressly promotes the idea of ASBOs as capable of enabling a *positive* process of engagement among local authorities, the police, housing professionals and residents. Rather than presenting a wholly critical perspective on anti-social behaviour policy as borne out of politically opportunistic and reactionary motivations situated within the modern discourse of popular punitivism (for further discussion, see Garland, 2001), this book contends that the creation of the ASBO, and latterly the introduction of the New Labour administration's 'Respect Agenda', in fact represent a socially *progressive* attempt to address the pernicious and debilitating effects of anti-social behaviour which are most often felt by those with the least opportunity to escape it (Millie et al., 2005).

As we shall see in the chapters that follow, there are (a small number of) scholars who have sought to challenge the overarching academic consensus on anti-social behaviour policy and the use of ASBOs. However, there are also, in my opinion, other examples of academic writing on the subject which go beyond what is a balanced, objective and *fair* analysis of the available evidence. Of course, as others have argued, philosophical and theoretical leanings/persuasions are entirely appropriate within the context of an opinion-based piece of work (see for example, Braswell and Whitehead's (1999) insightful analysis of 'truth-seeking' in criminology). Indeed, I am not arguing for a vacuum in judgment, opinion, or theorem in academic writing on ASBOs. It is axiomatic that academics possess theoretical biases which may, for example, take them in particular directions with regard to their research and naturally, theoretical biases will also inform any opinion-based piece that they write. The difficulty that exists with anti-social behaviour policy, I would argue, concerns (1) how research evidence is analysed in the context of theoretical biases and (2) how research and opinion is represented to students.

I have argued elsewhere that much of the available literature on anti-social behaviour policy and the use of ASBOs has necessarily been impaired by too much ideological partisanship (Donoghue, 2008). Essentially, what I believe is missing from current scholarship is a sense of *balance* in the debate. The advancement of academic thought in this area seems both implacable and relentless in its sustained critique of ASBOs and the New Labour administration's policy on anti-social behaviour. Perhaps this sustained critique is because – as one criminologist has suggested to me – there is 'simply nothing good to say about the ASBO'. And of course, this is certainly a popular point of view. Indeed, this book acknowledges that there are inherent problems associated with the ASBO model. In the pages that follow, we will see that although the academic critique of ASBOs is diverse (ranging from policy to implementation to ideology), there are essentially three broad strands of criticism levelled at the use of anti-social behaviour orders that this text will consider: first, that ASBO interventions are frequently inconsistently applied and *disproportionately* intrusive (Burney, 2002; 2005); second, that the use of ASBOs does nothing to tackle the underlying causes of anti-social behaviour and fails to rehabilitate those who are given ASBO interventions (Carr and Cowan, 2006; Scraton, 2004); and third, that ASBOs do not pay enough attention to due process and are unfairly targeted (primarily) at 'marginalised groups' such as young people and social housing tenants (Brown, 2004; Hester, 2000). While it is acknowledged that there are multiple vignettes to the ASBO critique, and so the above categorisation represents a general conflation of these arguments, it is nonetheless the position of this book that the academic view of ASBOs is pervasively critical.

Understandably, from a reading of current academic literature on anti-social behaviour policy, one would be forgiven for thinking that indeed there was 'nothing good to say about the ASBO'. But this presents a skewed picture of how the ASBO operates in practice, and moreover, the principles behind the ASBO model. My view remains that the result of the dominant academic view on anti-social behaviour has been to produce 'an imbalance in academic thought, whereby anti-social behaviour legislation and policy are too often discussed predominantly (or exclusively) in pejorative terms of (social) control. The subject of analysis has been, to a large extent, the *culture* of crime (and disorder) control and not an analysis of institutions and/or practice' (Donoghue, 2008: 338). In other words, while scholars of criminology and sociology have often sought to expound an ASBO culture of control embodying popular punitivism and 'responsibilisation' (Garland, 2001), and the identification of 'otherness' (Garland, 2001; Vaughn, 2000) what has been neglected in current scholarship is a more complete account of how ASBOs operate in practice, specifically with regard to how the institutions and agencies governing the use of ASBOs are operating and, moreover, what the measurable impact of anti-social behaviour policy and the use of the orders is upon victims and perpetrators of anti-social behaviour. Although this book concedes that there is currently the potential for ASBOs to be used inappropriately and to be disproportionately constraining, my argument is that there is also the opportunity for ASBOs to be used to enable a *positive* process of social protection – particularly for those with the least resources and opportunity to escape the worst effects of anti-social behaviour.

While an overarching critical academic view on anti-social behaviour orders is not necessarily problematic in and of itself, what I believe is of concern, is the unilateral way in which academia consequently represents anti-social behaviour policy and the use of ASBOs to students of criminology and sociology. Few course texts attempt to embody or to acknowledge alternative (theoretical) accounts of anti-social behaviour policy or, indeed, ASBOs. The result is that students are presented with a monolithic, sceptical academic perspective on anti-social behaviour management which, most frequently, has an undiluted focus upon analysing the control and management of newly-defined anti-social behaviour(s) through micro/macro management procedures, law, policy and other forms of bureaucratic control. There is then a corresponding inference that, for assessed work or exam answers, representation of and agreement with the dominant academic view is expected. Moreover, a unilateral representation of the 'social control' perspective on ASBOs is also reductive in terms of developing students' skills and confidence in being able to engage with and/or to challenge (in a scholarly fashion) argument(s) about anti-social behaviour management. It may be, of course, that most students are in agreement with the dominant academic view. However, alternative perspectives must be presented and engaged with to ensure that arguments made and conclusions drawn are rigorous and well-founded, and above all, that students are not stifled by the dominance of obliquity.

Undoubtedly, choices about what is taught, and *the way* that it is taught, are a substantive reflection of the values and principles of those in possession of the authority governing those choices. While it is accepted here – unequivocally – that value-freedom is not appropriate within all facets of sociology (indeed, as has already been alluded to, those works which seek to progress a specific orientation or argument are necessarily underpinned by value judgements) I believe that it is necessarily problematic to teach sociology from within a context which is predominantly – or exclusively – value-driven by those who teach it. This is not to say that those who teach students about aspects of anti-social behaviour policy ought to disguise or to subvert their own ideological or theoretical biases but rather, it is to suggest that pedagogy should strive for balance in the way that it selects and addresses topics for study, and in its awareness of the ideological content of its subject matter. As Kania (1988: 83) observes:

> The very act of selecting topics and issues in a course betrays one's social values, biases and prejudices; as does the failure to raise

issues. It is easier to see the bias in other points of view than our own. ... Approaching issues from a self-aware ideological perspective, confronting the opposing ideologies directly, has the advantage of exposing the biases of both points of view and allowing a truer examination of empirical reality than a single perspective ever could.

In this book, I have tried to present an alternative view on ASBOs which is (undoubtedly) critical of other scholarship but which nonetheless wholly (and *explicitly*) recognises the legitimacy of these arguments within the academic debate. It has been suggested that my concern about the dominance of academic critique on ASBOs represents little more than a discussion about a 'storm in a tea cup'. This, I feel, does a disservice to students if nothing else. Students should have access to all legitimate perspectives on anti-social behaviour policy in the course of their study. As such, there is a need for more diversity in points of view. I think that there is a need to consider whether it is possible to view ASBOs from a different theoretical perspective. To state it simplistically: can we conceive of ASBOs *beyond* negative social control? Rather than locating anti-social behaviour policy and the use of ASBOs (and indeed other anti-social behaviour interventions such as Acceptable Behaviour Contracts, Individual Support Orders and Parenting Orders) unilaterally within a prescriptive and exclusionary system of punitive social control, I will be looking at the reasons why the policy can, alternatively, be seen as forming part of process of social aegis. In this sense, the introduction of ASBOs can be characterised not necessarily as part of a 'culture of control' but as part of an ineluctable shift towards a culture petitioning civic reciprocity. In this regard, anti-social behaviour policy is underpinned by the principles of mutualism, and – most importantly – by the notion of a link between behavioural conduct and individual choice.

However, while I am broadly supportive of the ASBO model and of the intention behind it, (as I have already alluded to) this book does explicitly recognise that the ASBO system has a number of associated difficulties. In the pages that follow, drawing upon my own research on legal and court procedures, and other empirical work on the use of ASBOs, I advocate that the ASBO process be refined and ameliorated in order to address some of the problems associated with the ASBO model. Moreover, there is (as I, and others, have previously identified) a need for further research on ASBOs and anti-social behaviour management more generally. In 2005, the Home Office introduced 157 specialist

Anti-social Behaviour Response Courts situated within 31 Criminal Justice Areas in England and Wales, together with a network of 14 specialist anti-social behaviour prosecutors. The Economic and Social Research Council (ESRC) has kindly supplied me with a grant to study the operation of these specialist Courts, the findings of which will be available in 2011. The research will provide original data on the operation of these courts (about which there is currently no available empirical evidence) and which I hope will further elucidate how anti-social behaviour policy is working in practice, and to what extent ASBOs, and other anti-social behaviour interventions, are successfully 'protecting the community'.

But to return to the task in hand, it is my intention to highlight the existence of a prevailing intellectual climate which locates the 'social control' element as being of fundamental importance in understanding the operation and outcomes of current policy. Acting (in some instances tacitly, and in some instances expressly) as a means for 'exclusion', 'domination', 'demonisation', 'marginalisation' and 'stratification', social control has been identified as central to the construction of academic analyses of anti-social behaviour management and the use of ASBOs. Indeed, I will argue that the sceptical academic disposition towards anti-social behaviour management is characteristic of a more pervasive, overarching suspicion within sociology that sovereign power is generally to be viewed as a dangerous and clandestine entity vicariously linked to oppression and marginalisation (Loader and Walker, 2007). This dominant predisposition towards a scepticism about strategies of control means that, derivatively, academic perspectives on anti-social behaviour management are often imbued with a preconception that strategies of control are analogous to – and necessarily intersect with – notions of domination and subjugation. Of course, control *is* axiomatic in anti-social behaviour policy. However, I oppose the view that the only (or even the primary) focus of this control is the oppression and subjugation of the behaviour and expression of marginalised social groups. It will be argued that the current approach to anti-social behaviour should be seen as an amalgamation of separate and distinct elements; some undoubtedly related to forms of (positive and negative) control and some unrelated to control but which *as a whole* make up the contemporary paradigm of anti-social behaviour management. In other words, the study of ASBOs and anti-social behaviour management more broadly, should be concerned with research and writing about marginalisation, exclusion and intolerance *as much as* it should also be concerned with understanding the potential positive impacts for social cohesion, security and social capital. However, that ASBOs are part of a punitive culture of control which deliberately seeks to marginalise and to exclude 'targeted populations' is an argument which this book fundamentally rejects.

2
The End of Respect?

During the 2005 general election campaign, the Prime Minister Tony Blair began to promote an approach to urban disorder that was designed to put 'the law-abiding majority back in charge of their local communities'. Through the use of anti-social behaviour strategies and interventions, Tony Blair appealed to popular notions of civic responsibility and stated: 'I want to make [anti-social behaviour] a particular priority for this government, how we bring back a proper sense of respect in our schools, in our communities, in our towns and our villages'. Aimed at improving community cohesion by tackling anti-social behaviour 'more effectively', the policy was officially launched as the 'Respect Agenda' on 10 January 2006. However, shortly after Gordon Brown became Labour Party leader and Prime Minister in June 2007, the Respect Agenda, which had been one of Tony Blair's flagship policies during his term in office, was effectively sidelined. The drive to address anti-social behaviour through the use of ASBOs (amongst other new interventions such as Parenting Orders and Closure Notices) coupled with New Labour's broader political campaign on community values and 'rights and responsibilities', appeared to have run out of steam. Gordon Brown had split responsibility for anti-social behaviour policy between the Home Office and the new Department for Children, Schools and Families (DfCSF). Responsibility for youth justice was divided between the DfCSF, headed by Ed Balls and the Ministry of Justice. The popularity of ASBOs seemed to be seriously dwindling, with the number of new orders issued showing a decline of 34 per cent between 2005 and 2006, while Ed Balls proclaimed in July 2007 that he wanted 'to live in the kind of society that puts ASBOs behind us'.

By October 2007, it was announced that the Respect Task Force was to be replaced with a Youth Taskforce to support local delivery of the government's strategy on anti-social behaviour, with a particular focus on promoting earlier intervention and support to young people. Gordon Brown's decision to close down the Respect Task Force – which was based at the Home Office and which was responsible for co-ordinating the government's policy on anti-social behaviour – was seen by some commentators as signalling the end of the Respect Agenda. Moreover, it was suggested that the Respect Taskforce, with its high-profile director Louise Casey, had embodied an emphasis on anti-social behaviour that had served to reproduce a negative and misrepresentative image of British young people. It was subsequently argued that replacing the Task Force was a conscious choice by the new Prime Minister to move away from the anti-social behaviour/youth disorder rhetoric associated with his predecessor Tony Blair. One shadow minister went as far as to suggest that: 'The government has just airbrushed its whole Respect agenda out of existence' (*The Guardian*, 24 December 2007).

By dividing up responsibility for youth justice and anti-social behaviour, and replacing the Respect Task Force, Gordon Brown appeared to have relegated anti-social behaviour as a political priority. This led criminologists to wonder what direction the apparently downgraded policy would now take, and, furthermore, if indeed it would survive much longer in its new, seemingly demoted, policy incarnation. It was suggested that Gordon Brown's changes could infer signs 'that the government may be softening its aggressive rhetoric on anti-social behaviour and youth' (Squires, 2008: 369). Additionally, in the wake of the global economic crisis, it seemed as though there might be more immediate matters for the government to prioritise. In a country where rising unemployment and economic hardship had become salient political issues, it indeed seemed unlikely that the government would be interested in re-asserting its commitments to tackling anti-social behaviour in such a somber economic climate where peoples' immediate concerns seemed to lie elsewhere. Yet this is in fact, exactly what happened.

A 'new drive' on anti-social behaviour

In June 2009, Alan Johnson became Home Secretary following a cabinet reshuffle. Having grown up on a number of large housing estates, Johnson, like a number of Home Secretaries before him, drew upon his own experiences of anti-social behaviour to elucidate his policy perspective. For example, while Jack Straw (Home Secretary 1997–2001, and a key proponent of Blair's anti-social behaviour policy) had described living in a council estate with his mother, who experienced neighbour difficulties, Johnson spoke about the way in which residents of housing estates can feel 'ignored' and 'trapped in their homes'. It has since become evident that Johnson views anti-social behaviour policy as an important feature of New Labour's political agenda. Soon after becoming Home Secretary, Johnson stated that he wanted to ensure that criminal justice policy embodied a specific emphasis upon anti-social behaviour. In July 2009, he argued that:

> The focus must be on listening to the public, looking at what practical steps need to be taken to make the current system, with all the powers and responsibilities that this government has introduced, respond to their concerns…Worry about crime is seriously debilitating. If on some streets or estates, there are people who feel they can't step out after dark to buy a bottle of lemonade because they are fearful of the people they might find hanging around the stairwell or outside the off-licence, it has a profound impact on their life. (The Guardian, 2 July 2009)

Only a year before, it had seemed as though anti-social behaviour policy was being – not necessarily politically exiled – but certainly relegated in the political scheme of things. Some commentators had called it 'the end of the Respect Agenda'. However, this now appears to have been a rather premature conclusion.

The current Home Secretary has stated the government's renewed interest in anti-social behaviour policy and it is of particular significance that the Home Secretary has in the past been a key supporter and ally of Tony Blair having previously been referred to as a member of the 'Blairite' camp within the New Labour administration. During Tony Blair's time in office, Alan Johnson was a loyal proponent of New Labour policies on law and order and was associated with Blair's 'modernising' agenda for the Labour Party.[1] Following his appointment as Home Secretary, Johnson has been keen to revive, in particular, New Labour's policy on anti-social behaviour.[2] Indeed, in the statements that he has made since his appointment in June 2009, Johnson's rhetoric on crime and anti-social behaviour appears to mirror key elements of Tony Blair's early discourse on law order. In a speech that the Home Secretary delivered in July 2009, the law and order discourse of early New Labour was evident, as were statements alluding to those notions of 'community' and shared values originally developed as part of Tony Blair's Respect Agenda. Johnson argued for improved efficiency in the operation of anti-social behaviour interventions coupled with better systems of support since 'being tough on the causes of crime has been in many senses, the raison d'etre of this government over the last 12 years' (Home Office, 2009). The specific reference to Tony Blair's famous political slogan: 'tough on crime, tough on the causes of crime', is significant and demonstrates the continuity of policy: Johnson had aligned his policy perspective with Blair's original 'law and order' vision. Restating New Labour's commitment to addressing anti-social behaviour and its new 'drive' to tackle it as a priority, Johnson argued that:

> [T]he best way to restore public confidence ... is through tough action on anti-social behaviour. If the public sees that anti-social behaviour is being addressed, and if they appreciate that local councils are prioritising tackling graffiti, litter dropping, broken windows, abandoned cars – the kind of behaviour that has a profound impact on people's health and wellbeing – confidence and appreciation will multiply. Councils and the police now have more powers to deal with anti-social behaviour than ever before.... When these powers are used, they are highly effective. And they demonstrate clearly that the authorities are on the side of the public. Against those who seek to damage their communities and damage lives.

Johnson went on to say that despite a period of 'intense activity' in dealing with anti-social behaviour, the government had then unfortunately suffered an interim period of 'complacency' which required that attention once again be directed at anti-social behaviour policy and initiatives. Public anxiety about anti-social behaviour had not dissipated and the cases of Garry Newlove and Fiona Pilkington (which were widely reported in the media) had served to push anti-social behaviour further up the political (and media) agenda. In August 2007, Garry Newlove was kicked to death outside his home by a group of young people after he confronted them about damage to his wife's car and in October 2007, Fiona Pilkington killed herself and her disabled daughter after a decade of persistent abuse from young people living in her neighbourhood. Alan Johnson has recently stated that he intends to 'use the Fiona Pilkington case' to ensure that more is done by the police to tackle anti-social behaviour. Moreover, Gordon Brown appears to have made a strategic attempt to re-align himself with New Labour's 'rights and responsibilities' policy discourse (indeed, he specifically used this phrase in his 2009 Party Conference speech) and has proposed that new measures be introduced to ensure that the parents of any child guilty of anti-social behaviour will be automatically subject to a parenting order.

The renewed government interest in tackling community disorder focuses on 'graffiti, litter dropping, broken windows, abandoned cars – the kind of behaviour that has a profound impact on people's health and wellbeing'. The Ministry of Justice has announced the development of specific 'action areas' to assist with this task. The government's Anti-social Behaviour Action website is to be made more accessible to the general public, providing details on what action local authorities are taking on anti-social behaviour in comparison to other parts of the country. For those people who do not have access to the internet, leaflets are to be provided giving details of the number of ASBOs issued in their area, the numbers of crack house closures, and other details relating to anti-social behaviour management such as how many parenting orders have been granted. It is worth noting that the Home Office's official Anti-social Behaviour Action Website retains its conceptual link to the Respect Agenda in its web address (www.respect.gov. uk), and although responsibility for anti-social behaviour has now been split between the Home Office and the DfCSF, the priorities of the government's anti-social behaviour policy – although revised – continue to remain focused on reducing neighbour(hood)/community disorder and, in particular, youth-related anti-social behaviour.

Additionally, the processes by which anti-social behaviour is reported to the authorities is to be improved through better inter-agency co-ordination at a local level. According to the Home Secretary, at present, 'too many people who try to bring anti-social behaviour to the attention of the authorities find themselves trapped in a never-ending circle of phone calls – they phone the police, who tell them to phone the housing people, who tell them to phone social services, who tell them they need to talk to the police. Tackling anti-social behaviour effectively – particularly persistent offenders – depends on strong, local partnerships that have the expertise to address complex problems within communities' (ibid.). And as part of this new programme of measures to tackle anti-social behaviour more effectively – both in terms of outcomes and (importantly) in terms of the *speed* at which outcomes are delivered – the Home Office has also recently created the Anti-Social Behaviour Action Squad (ASBAS). The principal duties of the ASBAS is to set up local panels that will in turn provide guidance and support to frontline professionals to enable them to address different forms of anti-social behaviour more effectively, using the most appropriate interventions available. While the panels are in the first instance to be offered to those areas with the greatest public anxiety about anti-social behaviour, they are also to be made available in other areas where they may be of benefit.

Although there have been a number of difficulties associated with the processing of ASBO cases at magistrates' court, the length of time that it takes a case to come before the court – in some cases up to two years – has been specifically identified as requiring urgent attention. (The procedural and administrative difficulties presented by ASBO applications will be discussed in detail in Chapter 6.) As such, delays in waiting times before cases come to court are to be reduced and limits placed upon the number of times cases can be adjourned. Finally, given the psychological and emotional impact that anti-social behaviour can have on its victims, the Home Secretary has also raised the issue of how relevant agencies can best provide advice and prospective support to victims, stating that: 'there needs to be more support for victims who report anti-social behaviour, particularly those who have had to endure the most extreme forms of intimidation and harassment over many years' (ibid.). In turn, this necessarily raises questions about what form this support ought to take and whether it should be similar to the victim support services provided to victims of crime. In the proceeding chapters, we will consider more fully the (financial and emotional) costs of anti-social behaviour upon victims, and the difficulties that victims face during the ASBO court process.

These new objectives are significant developments in a policy that had been, by the Home Secretary's own admission, 'coasting' for the past couple of years. But the introduction of new strategies and initiatives must be understood in the context of how anti-social behaviour interventions have been working up until this point. In the pages that follow, we will examine the operation and effect of anti-social behaviour policy (in particular, the use of ASBOs), together with empirical evidence on how, and against whom, ASBOs are being used in Britain. However, for now, it is enough simply to conclude that despite Gordon Brown replacing the Respect Task Force and splitting responsibility for anti-social behaviour policy between the Home Office and the new DfCSF, anti-social behaviour clearly remains an important policy issue. And it is fair to say that the recent resurrection of anti-social behaviour policy is in large part a result of Alan Johnson's appointment as Home Secretary. While the Prime Minister Gordon Brown has been relatively quiet on the issue, it has been Johnson who has sought to refocus attention on anti-social behaviour as a significant political priority. Described as being in possession of a 'perfect working class

background' (Johnson is an ex-postman), coupled with 'Blairite tenden-
cies', the Home Secretary appears to be politically well-placed to revive
and to continue a policy that originated more than a decade ago under
Tony Blair's leadership. Whether Johnson will leave a personal imprint
upon the policy by moulding it in one direction or another, remains to
be seen. It will, however, be interesting to observe how the new 'drive'
on anti-social behaviour operates in practice, and what new directions
anti-social behaviour management will take in the near future. For the
meantime, it appears that it is not 'the end of Respect'.

3
Anti-Social Behaviour: The Political Landscape

Since its inception in 1997, the New Labour government has placed tackling anti-social behaviour and neighbourhood disorder at the forefront of its policy agenda. A substantial range of new legislation has been introduced, aimed at addressing 'selfish and unacceptable activity that can blight the quality of community life'[1] such as vandalism, graffiti, intimidation, harassment and noise nuisance. Moreover, between 2004 and 2008, the government launched two successive high-profile campaigns both with the objectives of raising awareness of the new measures created to tackle anti-social behaviour and improving their efficacy at local level. The 'Together' campaign (launched in 2003) was designed to improve local responses to tackling anti-social behaviour and latterly, the more widely recognised 'Respect Agenda' (launched in 2006) focused on the most effective measures for tackling anti-social behaviour but with a greater focus upon early intervention and support strategies for anti-social behaviour perpetrators.

Without doubt, the anti-social behaviour order (ASBO) is the most significant new power to be created as a result of New Labour's policy on anti-social behaviour, in part because of its hybrid civil law status (with criminal sanctions upon breach), but also by virtue of the subjective nature of the provisions and terminology embodied within the relevant legislation. The ASBO was first introduced in Britain in 1998 as a civil order, which was designed to protect individuals from acts of anti-social behaviour 'that cause, or are likely to cause, harassment, alarm or distress' under s.1(1)(a) of the Crime and Disorder Act 1998.[2] In England and Wales, an order can be made against anyone of 10 years old or over, although in Scotland, ASBOs were available only for persons aged 16 years old or over until a subsequent amendment in the Anti-Social Behaviour Etc. (Scotland) Act 2004 extended ASBOs to

12–15 year olds. An order contains conditions ('prohibitions') forbidding the offender from specific anti-social acts or entering defined areas and is effective for a minimum of two years.[3] The applicant agency must show that the defendant has behaved in an anti-social manner and that the order is necessary for the protection of persons from further anti-social behaviour by the defendant – this is sometimes referred to as the two-stage test. The agencies that are able to apply for orders are defined as 'relevant authorities' for the purposes of the legislation. In England and Wales, these are local authorities, police forces (including the British Transport Police), Registered Social Landlords (RSLs) and Housing Action Trusts (HATs). In Scotland, a relevant authority is a local authority, RSL or housing association. It is important to note that the police *cannot* apply for orders in Scotland, which makes ASBOs appear largely as a housing issue and goes some way to explaining the lack of private sector and owner occupation ASBOs in Scotland.

Interim orders are available under s.1D of the 1998 Act (as amended by s.65 of the Police Reform Act 2002) and s.7 of the 2004 Act in Scotland (as amended by s.44 of the Criminal Justice (Scotland) Act 2003). This temporary order can impose the same prohibitions and has the *same penalties* as breach of a full ASBO. Moreover, following legislative changes made in s.64 of the Police Reform Act 2002 and s.234AA of the Criminal Procedure (Scotland) Act 1995, criminal courts may now also make orders against individuals convicted of a criminal offence (sometimes referred to as a 'CrASBO'). In a similar way to ASBOs imposed in the civil courts, ASBOs on conviction are intended to prevent further anti-social behaviour, but specifically in relation to incidents that the police have reported (and where criminal proceedings have subsequently been taken). An order on conviction is granted on the basis of the evidence presented to the court during the criminal proceedings and any additional evidence provided to the court after the verdict.

As set out in the relevant legislation, applicant authorities have *a duty to consult* other agencies before an application for an anti-social behaviour order is made: Table 3.1 sets out the relevant statutory consultation requirements for applicant agencies. In addition, in Scotland, subsection 11 of the 2004 Act requires a relevant authority to consult, where the application relates to someone under 16, the Principal Reporter.

ASBO proceedings can be conducted in the magistrates' court (a stand-alone order can be obtained from the Magistrates' Court acting in its civil capacity); the Crown, Magistrates' or Youth Court (on conviction in criminal proceedings); or in the County Courts (orders can be made by a County Court where the principal proceedings involve

Table 3.1 Statutory consultation requirements

Relevant authority	Must also consult
Local authority	Police
Police (not Scotland)	Local authority
Registered Social Landlords (RSLs)/ Housing Action Trusts (HATs)	Police & local authority
British transport police (not Scotland)	Police & local authority

the anti-social behaviour of someone who is a party to those proceedings, although the court cannot make a stand-alone order as there must always be principal proceedings to which the application for an ASBO can be attached). In Scotland, application proceedings are heard in the Sheriff Court sitting in its civil capacity, or the Court of Session (appeal hearing), or the District Court or Sheriff Court on conviction in criminal proceedings.

Defining anti-social behaviour

The legal definition of anti-social behaviour that applies in relation to ASBOs, as conduct that: 'causes or is likely to cause harassment, alarm or distress to one or more persons not of the same household', inevitably and frequently results in a very broad range of behaviour falling within its scope. Although it is not an exhaustive list, the Home Office has, however, produced a typology of specific behaviours categorised as anti-social from a 'one day count' of anti-social behaviour carried out in 2003 (see Figure 3.1). Anti-social behaviour spans both criminal and non-criminal behaviour: subsequently, this has meant that behaviour complained of need not necessarily be of itself unlawful. The key feature of the statutory definition of anti-social behaviour is that its primary focus is the *effect* of the behaviour complained of: it is not necessary for the applicant authority to prove intention on the part of the defendant to cause harassment, alarm or distress.

While the relevant legislation defines anti-social behaviour broadly, and in terms of the *affective* consequences of the conduct, local authorities, RSLs, Crime and Disorder Reduction Partnerships (CDRPs) and police services possess their own lists of behaviours defined as anti-social for the purposes of the anti-social behaviour strategies within their locales. Relevant authorities have discretion over the creation of their own strategies to tackle anti-social behaviour[4] and these definitions

Figure 3.1   Typology of anti-social behaviours

Misuse of public space

- **Drug/substance misuse**
  - Taking drugs
  - Sniffing volatile substances
  - Discarding needles/drug paraphernalia
- **Drug dealing**
  - Crack houses
  - Presence of dealers/users
- **Street drinking**
- **Aggressive begging**
- **Prostitution**
  - Soliciting
  - Cards in phone boxes
  - Discarded condoms
- **Kerb crawling**
  - Loitering
  - Pestering residents
- **Illegal campsites**
- **Vehicle related nuisance**
  - Inconvenient/illegal parking
  - Car repairs on the street/in gardens
  - Abandoning cars
- **Sexual acts**
  - Inappropriate sexual conduct
  - Indecent exposure

Disregard for community/personal wellbeing

- **Noise**
  - Noisy neighbours
  - Noisy cars/motorbikes
  - Loud music
  - Alarms

Continued

Figure 3.1   Continued

- ○ Noise from pubs/clubs
- ○ Noise from business/industry

- **Rowdy behaviour**
  - ○ Shouting and swearing
  - ○ Fighting
  - ○ Drunken behaviour
  - ○ Hooliganism/loutish behaviour

- **Nuisance behaviour**
  - ○ Urinating in public
  - ○ Setting fires
  - ○ Inappropriate use of fireworks
  - ○ Throwing missiles
  - ○ Climbing on buildings
  - ○ Impeding access to communal areas
  - ○ Games in restricted/inappropriate areas
  - ○ Misuse of airguns
  - ○ Letting down tyres

- **Hoax calls**
  - ○ False calls to emergency services

- **Inappropriate vehicle use**
  - ○ Joyriding
  - ○ Racing cars
  - ○ Off-road motorcycling
  - ○ Cycling/skateboarding in pedestrian areas/footpaths

- **Animal related problems**
  - ○ Uncontrolled animals
  - ○ Dog fouling

Acts directed at people

- **Intimidation/harassment**
  - ○ Groups of individuals making threats
  - ○ Verbal abuse
  - ○ Bullying
  - ○ Following people

Continued

Figure 3.1 Continued

  ○ Pestering people
  ○ Voyeurism
  ○ Sending nasty/offensive letters
  ○ Obscene/nuisance phone calls
  ○ Menacing gestures
- **Can be on the grounds of:**
  ○ Race
  ○ Sexual orientation
  ○ Gender
  ○ Religion
  ○ Disability
  ○ Age

Environmental damage

- **Criminal damage/vandalism**
  ○ Graffiti
  ○ Damage to bus shelters
  ○ Damage to phone kiosks
  ○ Damage to street furniture
  ○ Damage to buildings
  ○ Damage to trees/plants/hedges
- **Litter/rubbish**
  ○ Dropping litter/chewing gum
  ○ Dumping rubbish (including in own garden)
  ○ Fly-tipping
  ○ Fly-posting

Source: (NAO, 2006: 39)

display a variance in the type(s) of behaviour(s) recognised as anti-social as a result of the differing cultural compositions and social conditions of specific locales. As a result, there is disparity and variational spread in the strategies employed to address anti-social behaviour(s) across Britain: anti-social behaviour legislation is thus underpinned by an emphasis upon local level autonomy.

The consequence(s) of such a generalised and subjective description of anti-social behaviour has meant that the definition is both flexible, and capable of diverse interpretation. Certain organisations have also commented that it has made anti-social behaviour more relevant and practical at a local level: the Local Government Association has argued that the 'anti-social behaviour focus from central government has led to an increase in focus within many localities' (House of Commons, 2004a: Ev. 81). Furthermore, Sergeant Paul Dunn of the Metropolitan Police has added that: 'the legal definition helps if enforcement is necessary, and it has to be looked at from that point of view' (House of Commons, 2004b: Q.96).

The Home Affairs Committee, reporting on anti-social behaviour in April 2005, made three main points relating to its wide definition:

> ...first, the definitions work well from an enforcement point of view and no significant practical problems appear to have been encountered; second, exhaustive lists of behaviour considered antisocial by central government would be unworkable and anomalous; third, antisocial behaviour is inherently a local problem and falls to be defined at a local level. It is a major strength of the current statutory definitions of antisocial behaviour that they are flexible enough to accommodate this. (House of Commons, 2005a: 21)

Alternatively, however, there exists a significant degree of opposition to such a subjective definition in law. For example, Hull City Council has argued that 'a lack of clarity around the definition of anti-social behaviour does not help' in producing an effective response to it, and Salford City Council has also highlighted this area as problematic (House of Commons, 2004a: Ev. 67 and 128, respectively). Moreover, the EU Commissioner for Human Rights, having examined the use of ASBOs in Britain in 2005, commented in his report that 'the determination of what constitutes anti-social behaviour becomes conditional upon the subjective views of any given collective' (Gil-Robles, 2005: 34).

While Scott and Parkey (1998) have suggested a threefold classification of anti-social behaviour to include: (a) neighbour nuisance, (b) neighbourhood nuisance and (c) crime, as ASBOs have become more widespread, it has been argued that the courts have become bolder and more inventive about how to frame such orders. Since anti-social behaviour relates to both criminal and non-criminal behaviour, a further consequence of the generalised nature of anti-social behaviour as behaviour that 'is likely to cause harassment, alarm or distress' has been that there is a clear diversity in the types of act that ASBOs can be granted to prohibit. The EU Commissioner for Human Rights noted: 'such orders [ASBOs] look rather like personalised penal codes, where non-criminal behaviour becomes criminal for individuals who have incurred the wrath of the community' (Gil-Robles, 2005: 34, para. 110).

As such an expansive scope of behaviour is necessarily open to 'social judgement', it has been suggested that the broad discretion available to applicant authorities has led to the targeting of those persons deemed 'undesirable', who, for the most part, possess marginalised and/or dis-advantaged social status. By way of example, since their introduction in April 1999, ASBOs have been used in certain circumstances, against the mentally ill, children with learning difficulties, the homeless, peace-ful protesters and prostitutes. This has met with criticism from a wide range of charitable and civil liberties organisations. For instance, The Children's Society has argued that 'many children and young people are telling us that they do not understand the term [anti-social behav-iour], but they feel it is directed towards them' (House of Commons, 2004a: Ev. 25). Crisis has also objected to the equating of begging as anti-social arguing that 'although the act of begging may be deemed anti-social, it is a problem that is best understood and dealt with as a manifestation of social exclusion' (House of Commons, 2004a: Ev. 37).

Hence, it is clear that there exists disparity among commentators, policy-makers and practitioners as to the effectiveness and value of having such a flexible statutory definition of anti-social behaviour that applies in relation to ASBO cases. Macdonald (2003) has observed that criticisms relating to the broad definition of the term 'anti-social behav-iour' possess a commonality centred on wider concerns about the place of discretionary autonomy within the legal system.[5] Thus, discretionary autonomy is a particularly important aspect of the administration and management of ASBOs. As Macdonald correctly identifies, the nature of the 'umbrella' term anti-social behaviour *necessarily precludes a con-cise, narrow definition*. Hence both local level governance/autonomy and judicial decision-making assume highly significant roles in decid-ing ASBO application outcomes (and the breadth and terms of ASBO prohibitions).

Incidence of anti-social behaviour

It is important to note that despite a number of studies having been conducted on the incidence of anti-social behaviour, there is a lack of consensus as to the proportion and extent of anti-social behaviour in

Britain. In Scotland, although overall crime (including serious violent crime) fell in the ten years to 2002, recorded offences of an 'anti-social' nature in Scotland increased (Scottish Executive, 2003). According to the Scottish Executive, these findings in fact understate the extent of the problem as much of anti-social behaviour is not within the criminal law, and goes unreported. Moreover, reports of vandalism in Scotland increased by almost 50 per cent in less than a decade: there were nearly 330 incidents a day reported on average in 2005–6, an increase from around 220 in 1996–7 (Scottish Parliamentary Written Answer, 2 March 2007). In total, there were 120,342 cases of vandalism, reckless damage and malicious mischief recorded in 2005–6, compared with 81,587 in 1996–7. Nonetheless, the extent to which an increase in reported anti-social incidents is equivalent to an increase in de facto anti-social behaviour is not comprehensively evidenced. Recent research has found that individuals in Scotland are now more likely to report incidents of anti-social behaviour to the authorities (Scottish Executive, 2005b). Pawson and McKenzie (2006) suggest that a greater willingness on the part of victims to report anti-social behaviour has come about as a direct result of increased media attention and the belief that authorities will now act positively to resolve the problem. Commentators have further observed that the increased (and increasing) attention and time devoted to reporting incidents deemed 'anti-social' within the media, has also consequently served to encourage public perception of the overall incidence of that type of behaviour in Britain.[6]

In England and Wales, despite a 39 per cent drop in the incidence of crime since 1995, anti-social behaviour continues to remain an issue of public concern with around 66,000 reports of anti-social behaviour made to authorities each day (Home Office, 2003b). Moreover, the Home Office estimates that around 17 per cent of the total population (approximately 7 million people) perceive there to be 'high levels' of anti-social behaviour in their area (Home Office, 2006a). However, the number of people who think that anti-social behaviour is a 'big' or 'fairly big' problem reduced from 20.7 per cent in 2002/03 to 16.7 per cent at the end of 2004 (Home Office, 2004b). Burney (2005: 60) has argued, however, that because people living in different locales identify different types of behaviour as being more problematic than others,[7] that these variations in fact negate the effectiveness of having a wide definition of anti-social behaviour because of its illogical categorisation of such widely varied phenomena within one annotation. Research evidence also shows that specific groups of people are more likely to be affected by anti-social behaviour than others. For example, 30 per cent of those living in social housing and 32 per cent of those living in 'hard pressed' areas – who are least able to move away or bear the cost of anti-social behaviour – perceived high levels of anti-social behaviour in their area. Similarly, findings show that those individuals from an ethnic minority (26 per cent) and females aged between 16 and 24 (28 per cent) found anti-social behaviour to be a 'big problem' for their area (NAO, 2006: 9). (In Scotland, similar findings have been reported: Scottish Executive *Guidance on Antisocial Behaviour Strategies* (2004a) notes that, in particular, vulnerable groups such as older people, women and disabled people, including children/adults with mental health or learning difficulties, are likely to be more affected by anti-social behaviour and the fear of crime.)

Age breakdown is an important factor in determining the incidence of anti-social behaviour in relevant areas. In Scotland, the findings of the Scottish Crime Survey 2000 (Scottish Executive, 2001) and the Scottish Household Survey 2001/02 (Scottish Executive, 2002) demonstrated that the most commonly cited 'neighbourhood problem' was 'groups of young people [hanging around]'. Nearly a third of all respondents cited this as 'fairly or very common in my area'. Similarly, evidence from research in England and Wales also shows that youth disorder is a primary concern of individuals within the typology of anti-social behaviour(s) (Home Office, 2006a). Hence, if 'groups of young people [hanging around]' is the primary concern of residents then this could be a factor in determining a perceived higher incidence of anti-social behaviour, especially if an authority has in the past been guilty of re-housing 'problem families' in certain estates.[8] Evidence has shown that there can often be a larger proportion of young people living in social housing areas than in other tenures, perhaps increasing pressure on local facilities and leading to a perception by residents and agencies of increased anti-social behaviour (see, for example, SEU, 2000a). Thus, there is a lack of consensus as to the level and prevalence of anti-social behaviour in Britain. The incidence – and perception – of anti-social behaviour is subject to variation as a result of differences in geographical area and housing tenure, and by virtue of the gender, age and ethnicity of anti-social behaviour perpetrators/victims. Moreover, the broad legal definition of the term 'anti-social behaviour' embodied within the relevant legislation also means that it is very difficult to *legitimately compare* the incidence of anti-social in recent years with historical evidence on disorder.

Frequency of ASBO use

In Scotland, between April 1999 and March 2005, 559 ASBOs had been granted (this includes those initially granted on an interim basis) (Scottish Executive, 2005b: 1). In the most recent study year (2004/05), a total of 205 ASBOs were granted by the Scottish courts, representing a rate of 9.2 Orders per 100,000 households, and represents a decrease in the rate of growth in ASBO activity in Scotland, which had been continuously increasing since 1999/00. Research on behalf of the Scottish Executive notes that the incidence of ASBO applications in Scotland 'is not only highly diverse, but is also quite inconsistent with what might be anticipated in terms of the expected pattern of anti-social behaviour' (Scottish Executive, 2005a: s.2.25). Moreover, ASBO activity is 'only slightly associated with survey evidence on the incidence of anti-social behaviour' in Scotland (Scottish Executive, 2005b: 1). The reasons for this geographical variation in ASBO use are subsequently described as fourfold (ibid: 2) and include the differing speeds at which local authorities/RSLs have been 'gearing up' to make full use of ASBO powers; the variation in attitudes of the legal profession/courts regarding ASBO applications; the organisational responsibility for tackling anti-social behaviour within individual local authorities; and the extent of local authority/RSL commitment to resolve anti-social behaviour through alternative means such as mediation, the use of anti-social behaviour contracts (ABCs) etc. About half of all full ASBOs granted in both 2003/04 and 2004/05 in Scotland were of indefinite duration (Scottish Executive, 2005a, s.5.5) and orders range in length from less than a year to an indefinite duration. For some landlords, placing an indefinite duration upon an ASBO is 'simply a standard approach or part of their official policy' (s.5.8).

There is also diversity in the types of act that ASBOs are being granted to prohibit, which is, in part, a consequence of the broad definition of anti-social behaviour as behaviour that causes, or is likely to cause, 'alarm or distress'. Research has found that, in Scotland, behaviour prompting an ASBO application generally falls into one of three main categories – neighbour nuisance, noise and rowdy behaviour. Of the total number of ASBOs granted in Scotland, 40 per cent relate to noise nuisance and a further 11 per cent of orders prohibit the perpetrator from entering a specified area. In terms of the conditions placed on ASBOs, 44 per cent were classed as 'other' by local authority/RSL respondents and involve the prohibition of a wide range of behaviours, including shouting, swearing, vandalism, verbal abuse, threatening behaviour, intimidation and carrying a weapon (Scottish Executive, 2005b: 3).

In England and Wales, the most recent figures show that the total number of ASBOs issued (at the end of December 2007) stood at 14,972 (Home Office, 2008). Research also demonstrates that the same factors affecting the wide geographical variation in the use of orders in Scotland are affecting the issuing of orders in England and Wales. Both Campbell's early work on the use of ASBOs and the most recent study of the orders carried out by the National Audit Office similarly highlight geographical variation and the attitudes of practitioners to anti-social behaviour interventions (and available alternatives) within different locales as effecting ASBO uptake (see Campbell, 2002; NAO, 2006).

The cost of taking out an ASBO has been known to range from £2500 to in excess of £100,000.[9] Campbell's (2002) early review of the financial costs associated with ASBOs concluded that the average cost to police or local authorities was £4800, including case preparation, attendance at problem-solving meetings and dealing with breaches and appeals. In 2004, the results of the Home Office's 'Together' *ASBO Cost Survey* then provided an estimate of £2500 for the average cost of obtaining an ASBO. These results suggest that the average cost of administering an ASBO has fallen since Campbell's analysis in 2002. The report notes that 'the main drivers behind this decrease in costs appear to be the use of ASBOs on conviction and possibly more efficient administrative and legal procedures, as practitioners have become increasingly familiar with using ASBOs' (Home Office, 2004c: 2). However, the report findings add that estimates of the costs of ASBOs were wide-ranging in both the 2002 and the 2004 surveys.[10] Alternatively, many local authority anti-social behaviour protocols state that the minimum cost of an ASBO application is likely to be £5000 and will involve several weeks/months of preparatory work. However, the nature of ASBO applications, the diversity of those made subject to them and the differences between the authorities applying for them mean that no 'standard cost' of an ASBO application can be given.[11]

Effectiveness

Until very recently, there has been a dominant focus upon the bureaucratic aspects of the use of ASBOs in existing empirical research, and an absence of evaluations of 'what works' in reducing anti-social behaviour. For example, the Scottish Executive's most recent research project on the use of ASBOs (Scottish Executive, 2005a; 2005b) was concerned with 'monitoring' the use of the orders – it was not concerned with evaluating or analysing effectiveness, or attempting to determine quantifiable 'successes' and 'failures'. As research to date has largely been concerned with investigating the administration and application of the orders, a large proportion of the evidence on the effectiveness of ASBOs has been essentially anecdotal. Burney (2002: 481) found that there exist anecdotal examples of a reduction in anti-social behaviour as a result of certain orders being granted, and similarly, examples of the ineffectiveness of the orders, but 'no means of knowing whether they add up to a significant whole.' Moreover, the few evaluations that are in existence have been carried out locally and with very little standardisation in methodology (Armitage, 2002).

A report published on behalf of the National Audit Office in December 2006 was, however, the first national study (in England) to attempt to review the use of ASBOs with other anti-social behaviour interventions (warning letters and acceptable behaviour contracts (ABCs)), with the purpose of providing an analysis of whether interventions were successful in deterring further acts of anti-social behaviour. The study sampled 893 case files of ASBOs, warning lettersand ABCs issued in six areas: Wear Valley, Easington, Liverpool, Manchester, Exeter and Hackney. The study found that (in the cases sampled by the Audit Office) almost two-thirds (65 per cent) of people stopped behaving anti-socially after one intervention; over four out of five stopped after two interventions; and after three interventions, anti-social behaviour had been stopped in more than nine out of ten cases. A small proportion of individuals were, however, repeatedly engaged in anti-social behaviour. The report found that approximately 20 per cent of the sample cases received a (disproportionate) number of interventions – totalling 55 per cent of all interventions issued in the period covered by the study (NAO, 2006: 5). This same group had a higher number of average convictions (50) than those in the total study sample who also possessed convictions (24) (ibid.). The report also suggested that about 55 per cent of anti-social behaviour orders had been breached by offenders either committing more offences or by breaking the terms of their orders (NAO, 2006: 7). While the average number of breaches was four per person, the report found that 35 per cent of ASBO holders breached their orders on five or more occasions (ibid.).

The government has argued that, where breaches are reported it means that individuals are being monitored and that communities feel confident enough to report them. Alternatively, critics contend that the orders can only be effective if they are properly enforced, and that the existence of the figures on breach demonstrates that this is not the case. However, some practitioners have maintained that breach of an order 'is not necessarily a failure'. It may be the case that the terms of orders are breached on more than one occasion before the behaviour stops. That the order was breached does not necessarily equate to de facto failure if the intervention does succeed in stopping (or even reducing) the anti-social behaviour. As one anti-social behaviour unit manager has argued:

> People who are very critical of ASBOs give the fact they have been breached as evidence of their failure. These are people who don't look in depth at the way ASBOs are working... people say that ASBOs fail because they are breached, which is absolute nonsense. We've got one where the police described the offending level as 70% of their work load in the area. It's now been breached, it's been breached about twice, and really the police are saying this is remarkable, what's happened with these youths. They are off the richter scale as a problem now. The problem behaviour is so minimal now, compared to what it was. But if you see that as a statistic, that's a failed ASBO. It's really nonsense the statistics they're producing at the moment.[12]

Moreover, definitions of ASBO 'effectiveness' are necessarily contentious. While the National Audit Office study defined effectiveness as whether or not interventions were successful in deterring anti-social behaviour, opponents of the ASBO model contend that such a definition is over-simplistic and fails to take account of the individual circumstances of each case, and indeed, whether punitive action in the form of an ASBO is necessarily the most appropriate solution, particularly where anti-social behaviour perpetrators may already be socially excluded/marginalised, and might also have existing mental health or addiction problems. Issues relating to the disadvantaged status of anti-social behaviour perpetrators, and their responsibility for anti-social conduct, will be considered further in Chapter 6. However, the results of the National Audit Office's study does seem to suggest that anti-social behaviour interventions are, in the majority of cases, successful in deterring anti-social behaviour (although more than one intervention may be required before a person stops behaving anti-socially). Indeed, the Home Secretary Alan Johnson has, as we have seen, described the use of anti-social behaviour interventions as 'highly effective' and has introduced new policy objectives aimed at speeding up the ASBO process, making it more efficient, and providing better support for victims of anti-social behaviour.

Hence a key distinguishing feature of the New Labour administration in Britain has undoubtedly been its approach to crime and disorder control – and in particular, its controversial policy on tackling anti-social behaviour and the subsequent introduction of ASBOs. However, contemporary policy on anti-social behaviour management must also be understood in terms of the political landscape unto which it was born. In this next section, we will consider the political background to the introduction of New Labour's policy on anti-social behaviour, and the evolution of criminal justice from penal welfarism, to the new politics of law and order (Garland, 2001).

Penal welfarism and the new politics of law and order

The politics of 'law and order' did not always feature so veritably in political discourse. Indeed, crime and criminal justice policy did not begin to acquire political cachet until the late 1970s. Tracing the growth of 'law and order' issues pertaining to crime, policing and criminal justice in the post-war era, Downes and Morgan (2007: 204) identify the existence of what they term a general 'bipartisan consensus on law and order', whereby political parties in post-war Britain were, for the most part, broadly in agreement about matters which were of greatest importance to the country at that time (economic reconstruction and the welfare state) and matters which, as a result of pragmatic exigency, were of secondary importance (law and order). As a product of Britain's post-war economic and social predicament, crime and criminal justice were not seen as the divisive, partisan political issues that they have now become.

For example, prior to their election victory in 1974, the Labour Party's manifesto had largely been concerned with issues relating to the economy, employment, industrial relations and the welfare state. Crime, and criminal justice, did not feature prominently. Nor was Labour's manifesto statement on law and order issues particularly combative or adversarial.

The only point at which the manifesto commitments appear to infer a partisan critique is with regard to the potential privatisation of the police service. However, following the election of the Thatcher government in 1979, law and order issues came to occupy a far more strategic locus in British politics. Suggesting that worsening crime figures could be attributed to Labour government inaction, the Conservative Party manifesto for the 1979 election argued that the Labour government had undermined respect for the rule of law. Identifying a growing threat to public security from crime and disorder, the Conservative Party promised to prioritise law and order, and to spend more on fighting crime.

The increasingly antagonistic, partisan approach to law and order adopted by the Conservatives during the 1970s (culminating in victory for the Party in the 1979 general election) was mirrored by an inability of the Labour Party and the Liberal Party (subsequently the Liberal Democrats) to convince the electorate that they were as 'strong' on law and order as the Conservatives. It became apparent that crime and criminal justice policy had become progressively more important to the electorate. As such, the main political parties moved away from their tacit bipartisan consensus on law and order and sought to define themselves as possessing a distinct and recognisable partisan stance on crime and criminal justice issues. As Downes and Morgan (2007: 204) observe, the Conservative Party's stance on law and order in the early to mid-1980s remained 'tough' and uncompromising, while Labour 'initially stuck to their 'one nation' position, focusing on 'healing the wounds' brought about by unemployment and the Conservative cuts in public expenditure ... By contrast the Liberals ... took a more radically reformist position, arguing for incorporation into domestic law of the European Convention on Human Rights, the creation of a Ministry of Justice, and local authority crime prevention units'. Hence, the main political parties had sought to differentiate themselves from each other in respect of law and order policy and had substantively moved away from the broadly consensual approach that was evident in the immediate post-war years. Nonetheless, another period of broad agreement about law and order policy was to follow.

Although the years immediately proceeding the Conservative Party election victory in 1979 had borne witness to a somewhat reactionary process of political positioning on law and order, by the mid-1980s the political climate on crime and criminal justice was again shifting. The main political parties had begun to realise that crime – and its causes – could not be understood or addressed unilaterally. As a result, a more nuanced approach to law and order began to develop which recognised the antecedent social character of crime. Broad agreement was evident with regard to support for, and improving the effectiveness of, the police service. Moreover, there was an overarching bipartisan consensus that supported the rationale that there should be increased sentences for persistent, serious offenders coupled with a reduction in custodial sentences for low-level, petty offenders through the proliferation and increased use of non-custodial interventions (Downes and Morgan, 2007). However, adversarial, partisan politics on law and order was soon to return.

During the 1990s, political culture evolved in a considerably more punitive direction (Burney, 2005: 17). Research had demonstrated that crime had risen significantly over the preceding decade. In 1980, the official number of recorded crimes stood at 2.5 million but by 1992 it had increased dramatically to 5.4 million (Home Office, 1998). Moreover, during the period 1989–91, some areas recorded increases in the rate of crime of more than 30 per cent. While official statistics should necessarily be treated with caution given the multifarious elements capable of influencing their composition (most notably, the 'dark figure' of unrecorded crime; manipulation of crime figures to suit police and/or Home Office interests; and trend variation in the methodologies and practices of recording and reporting crime) the evidence of victim surveys suggests that in the period 1989–1993, official statistics correspond closely to genuine increases in victimisation.

Rising public concern about crime proffered an opportunity for the Labour Party, in opposition, to capitalise on popular perceptions that Conservative policy on law and order was ineffective and had failed to prevent a continued increase in the level of recorded crime. In a politically pragmatic move, the Labour Party was able to exploit the apparent weaknesses in the Conservative Party's record on crime whilst identifying rising crime levels as explicitly connected to leniency in sentencing (Downes and Morgan, 2007). Taking full advantage of the Conservative Party's vulnerable position on crime, Tony Blair (as Shadow Home Secretary) argued: 'It is high time the Tories took crime as seriously as the public and, instead of a series of attempts to ward off bad headlines, they put forward a thought-out policy to fight crime' (*The Independent*, 29 April 1993).

Concerned about their popularity in the public opinion polls, the Conservatives mounted a counter-attack on Labour and attempted to realign themselves as the Party of law and order. Upon his appointment as Home Secretary in 1993, Michael Howard began a transformation of Conservative crime control policy, from a focus on 'just deserts' principles in sentencing (embodied within the Criminal Justice Act 1991 which had emphasised the importance of community-based sanctions as opposed to incarceration, and had aimed at the rehabilitation and reintegration of offenders into society) to a new punitive approach premised upon the notion that 'prison works'. The 'prison works' approach to crime was underpinned by the argument that de facto crime levels could be reduced by the incarceration of greater numbers of offenders – particularly persistent offenders. On 6 October at the Conservative Party Conference, Michael Howard stated:

> Prison works. It ensures that we are protected from murderers, muggers and rapists – and it makes many who are tempted to commit

crime think twice.... This may mean that more people will go to prison. I do not flinch from that. We shall no longer judge the success of our system of justice by a fall in our prison population.

What followed was a record expansion of the prison population over the next four years (Morgan, 1997). By 1997, the Conservative Party's emphasis on increased incarceration, coupled with their advocacy of longer, 'harsher' sentences, had resulted in a landmark high in the prison population in England and Wales of 60,012 inmates. Rejecting criticism from penal reform campaigners, Michael Howard responded by arguing that the drop in recorded crime over the 4 preceding years, was directly linked to a rise in the prison population. Howard argued that incarcerating more offenders is effective because it stops criminals committing further offences and so reduces crime.

Downes and Morgan (2007: 215) have argued that both Conservative Home Secretaries after 1992 (Kenneth Clarke and Michael Howard, respectively) were *fundamentally* influenced in their policy-making decisions on law and order by their Labour Party shadow cabinet opposites (Tony Blair and Jack Straw as Shadow Home Secretaries). Indeed, they contend that 'Clarke and Howard [were] the prisoners of Blair and Straw's agenda, rather than – as is conventionally assumed – the reverse'. Alternatively, other commentators have sought to advance a perspective on Michael Howard's decision-making on crime policy as determined by his own sensibilities and attitudes to law and order, as opposed to a hastily formatted response to New Labour's attack. For example, in evidence to the Select Committee on Home Affairs, it was argued that: '[Michael Howard's] predecessors in office, void of their own ideas, were victims of a well established anti-prison lobby whose grip on Home Office policy-makers was absolute. In stark contrast, although always prepared to listen to sensible advice, Michael Howard had a mind of his own and made his own decisions' (House of Commons, 1998: Appendix K).

While the transformation of the Conservative Party's approach to law and order was undoubtedly motivated in part by the Labour Party's successful (if somewhat fortuitous) attack on the Conservative's record on crime, it is also worth noting that, out of cabinet office and out of government, Howard has consistently maintained his position on law and order and his conviction that '... an increase in the number of criminals in prison leads to a large fall in crime'. Indeed, Howard has in recent years also sought to revive the 'prison works' concept, despite the term being considered rather out of date now, not to mention the scepticism (and degree of ridicule) that the phrase often receives. Much crim-inological scholarship identifies Howard's 'prison works' as explained unilaterally by political opportunism. However, although a product of political pragmatism to a degree, Michael Howard's 'prison works' agenda can *also* be understood as more than simple opportunism and as projecting directly from his own beliefs about crime and his view on how it best be prevented.

In a similar vein, Jones and Newburn (2007) observe that Michael Howard's enthusiasm for mandatory sentencing as a means to deter and incapacitate recidivists was specifically borne out of his strong *personal* interest in the idea. His Crime (Sentences) Bill sought to introduce auto-matic life sentences for second time rapists or violent offenders and mandatory minimum sentences for both a third domestic burglary and also for a third Class A drug trafficking offence (New Labour subse-quently enacted the Crime (Sentences) Act 1997 embodying these man-datory sentencing provisions). While acknowledging that mandatory sentencing had existed in the US for several years before (most notably in California which had adopted the 'Three Strikes' policy of manda-tory sentencing in 1994), Jones and Newburn reject the proposition that policy transfer was particularly influential in Michael Howard's decision to adopt such a policy approach. As Simon (2008: 252) con-cludes: 'Howard was not ignorant of US proposals like "Three-Strikes" or "Truth-in-Sentencing", but he was thinking along the same lines from a common intellectual root, rather than borrowing or lesson drawing.'

But the Conservative Party's toughening line on law and order is only part of the explanation for the increasing punitive turn that crime pol-icy was taking during the 1990s. Of pivotal importance for the develop-ment of law and order policy was undoubtedly the metamorphosis of the Labour Party into 'New Labour' and the genesis of the Party's 'tough on crime, tough on the causes of crime' approach to law and order in Britain. In the aftermath of the Labour Party's fourth successive gen-eral election defeat in 1993, Tony Blair was elected as leader. Substantial changes to Party rhetoric – reflecting a move away from traditional social democratic principles – were immediately evident, as was internal party reorganisation and the creation of new policies that went beyond trad-itional Labour philosophical principles and which were capable of tran-scending the politics of the New Right associated with Thatcherism.

New Labour's (and, in particular, Tony Blair's) approach to 'law and order' was influenced by developments in the United States during the 1990s. In the previous three decades, the US had seen an upward trend in crime statistics. Street crime, in particular, had become a serious concern for communities in the US and fear of crime had intensified (Wilson, 1975). In turn, politicians in America had sought to demonstrate that they were tough on crime through revised law and order policies (see e.g., Ren et al., 2008). Social and political concern about rising crime consequently provoked a policy reaction aimed at finding a 'solution' to address 'the problem of crime'. At the start of the 1990s, crime remained on the increase. However, it then subsequently levelled off and then decreased considerably after 1994. During this period, Wilson and Kelling's (1982) 'Broken Windows' theory on crime and policing had become highly influential in American crime policy, particularly in New York.

Broken Windows

The 'Broken Windows' approach to crime essentially dictates that ignoring petty crime and disorder encourages criminal behaviour, eventually leading to more serious crimes. When low-level disorder is addressed, anti-social behaviour is deterred and (most contentiously) serious crime is reduced/prevented. Wilson and Kelling summarise as follows:

> We suggest that 'untended' behaviour…leads to the breakdown of community controls. A stable neighbourhood of families who care for their homes, mind each other's children, and confidently frown on unwanted intruders can change, in a few years or even a few months, to an inhospitable and frightening jungle. A piece of property is abandoned, weeds grow up, a window is smashed. Adults stop scolding rowdy children; the children, emboldened, become more rowdy. Families move out, unattached adults move in. Teenagers gather in front of the corner store. The merchant asks them to move; they refuse. Fights occur. Litter accumulates. People start drinking in front of the grocery; in time, an inebriate slumps to the sidewalk and is allowed to sleep it off. Pedestrians are approached by panhandlers. (1982: 32)

Moreover, the 'Broken Windows' theory is underpinned by the notion of 'zero-tolerance' policing towards petty crime and anti-social behaviour. The theory proposes that it is necessary to penalise every minor infringement of the law in order to create neighbourhoods in which serious crime will be less prevalent. Without this 'zero-tolerance' approach to low-level disorder

> …one broken window becomes many. The citizen who fears the ill-smelling drunk, the rowdy teenager, or the importuning beggar

is not merely expressing his distaste for unseemly behaviour; he is also giving voice to a bit of folk wisdom that happens to be a correct generalization – namely, that serious street crime flourishes in areas in which disorderly behaviour goes unchecked. The unchecked panhandler is, in effect, the first broken window. Muggers and robbers, whether opportunistic or professional, believe they reduce their chances of being caught or even identified if they operate on streets where potential victims are already intimidated by prevailing conditions. If the neighbourhood cannot keep a bothersome panhandler from annoying passersby, the thief may reason, it is even less likely to call the police to identify a potential mugger or to interfere if the mugging actually takes place. (p. 35)

While aspects of the 'Broken Windows' theory had been transplanted into New York crime policy since the mid-1980s, following his election as Mayor of New York City in 1993, the strategy of 'zero-tolerance' then became the cornerstone of Rudy Giuliani's crime policy reforms. Giuliani's zero-tolerance method of policing saw, for example, strict enforcement of the law against public drunkenness, vagrancy, subway fare evasion and 'squeegee men' (people who would wipe the windscreens of stationary cars without the driver's consent and then demand payment). This broadening of the focus of law and order policy towards incivility and anti-social behaviour embodied a zero-tolerance approach to disorder through punitive action against those who were, as Wilson and Kelling originally described, 'not necessarily, criminals, but disreputable or obstreperous or unpredictable people' (1982: 30). Crime rates dropped suddenly and significantly and continued to decrease over the next ten years (Kelling and Sousa, 2001): whether this fall in crime was a direct result of the 'Broken Windows' approach has been a matter of much debate and controversy. While the 'Broken Windows' thesis predicts a *causal* link between violent crime and disorder, other criminologists have observed that this relationship is in fact spurious (Sampson and Raudenbush, 2004). Moreover, critics have also sought to demonstrate that the fall in the rate of both petty and serious crime in New York during this period, was attributable to factors other than the zero-tolerance approach to policing (see, for example, Bowling, 1999; Karmen, 2001).

However, after his visit to New York in 1995 to observe and to bear witness to the apparently impressive outcomes of the zero-tolerance crime policy reforms, New Labour Shadow Home Secretary Jack Straw began to advocate the introduction of a similar approach in Britain.

While the main political parties' interest in mandatory sentencing policies was linked very tenuously with direct policy transfer, Jones and Newburn (2007) argue that 'zero-tolerance' was one policy approach that *was* strongly affiliated with direct policy transfer and convergence – although they contend that, through the introduction of ASBOs, '... the United Kingdom seems to have gone even further than the United States [by creating] civil measures that permit behaviour that is deemed disorderly and anti-social to be banned on pains of criminal punishment' (p. 132). Furthermore, Burney (2005) has argued that Jack Straw's enthusiasm for an approach to crime and disorder that was closely associated with the punitive, zero-tolerance US policy perspective resulted in New Labour's 'crack down' on incivilities, which came to be largely embodied within the Crime and Disorder Act 1998.

But the evolution of New Labour's law and order agenda had also been influenced by a form of communitarian ideology (Etzioni, 1993) which had grown popular in the United States during the early 1990s (although it had first originated as a theory during the 1970s). Drawing upon the notion of 'community', communitarian principles are concerned with 'community life' which embodies notions of responsibility and reciprocity between individuals based upon *shared values* of trust and respect (as we shall see in the pages that follow, the concept of 'respect' for one's neighbours became pivotal to New Labour's policy on anti-social behaviour). Most notable, perhaps, is the concept that rights entail responsibilities (Etzioni, 1993) which became the cornerstone tenet of New Labour's approach to community safety and anti-social behaviour management in the proceeding years. Reflecting upon New Labour's embodiment of communitarian principles into its policy on law and order, Calder (2003) observes that the communitarian vision of the 'community' is as

> ...something in need of enhancement or protection. It might be presented as a force for good – a counterbalance to factors perceived as threats to social cohesion and stability, from alienation or anomie among certain sectors of society, to the development of criminal subcultures or the growth in drugs misuse or teenage pregnancy. More generally, it might be invoked as the key to a rekindling of civic virtue, of a political culture which recognises the value of what individuals share, as well as what makes them different from one other.

And it was this notion of the 'community', and the enhancement of community life, that New Labour sought to invoke to guide its policy on law and order and, more generally, to guide the direction of New Labour and its neoteric political agenda. Instrumental to New Labour's policy approach was Giddens's conceptualisation of 'Third Way' politics (1994; 1998). As Tony Blair argued in a document published by the Fabian Society entitled *The Third Way: New Politics for the New Century* (1998), 'We all depend on collective goods for our independence; and all our lives are enriched – or impoverished – by the communities to which we belong. ... The life of any family and any community depends on accepting and discharging the formal and informal obligations we owe to each other. The politics of "us" rather than "me" demands an ethic of responsibility as well as rights.' In the next chapter we will consider in more detail the way in which the term 'community' has become a specific locus of governance, and moreover, the criticisms that have been made of how the term is differentially employed. However, at this juncture, it is important to note that by inculcating the notion of the primacy of the community, and *shared* community values, New Labour's 'Third Way' politics eschewed the individualism associated with the New Right – and thus Thatcher's observation that there was 'no such thing as society' – but it also spoke to traditional social democratic notions of social justice and the need for a strong state as the mechanism to achieve a greater degree of egalitarianism in society (Calder, 2003). Hence the ideology of the Third Way was distinct from both liberal capitalism – which supported free market economy principles (a market not burdened by government regulation/intervention) – and democratic socialism (with its prescriptive emphasis on the role of the state in achieving greater social equality). The Third Way endorsed economic enrichment, enterprise and wealth creation *as well as* greater social justice through processes of state intervention. Thus Third Way politics served to move the Party 'beyond Old Left and New Right', widening its appeal to the electorate substantially.

Tough on crime, tough on the causes of crime

Although it may seem on the face of it that New Labour's change of political direction towards a style of politics which was influenced by communitarian ideals would result in a genuine balance between control, discipline and social justice, critics have argued that New Labour's approach to law and order was soon to take an especially 'punitive' turn. Interestingly, in his progressive analysis of *The Politics of Law and Order* in the United States, Scheingold (1984: 204) speaks of the importance of 'symbolism' in politics – and what this in turn means for the adoption of a 'law and order' agenda. He identifies that individuals' perceptions about (the validity of) a given crime policy is not so much shaped by their own personal experiences/victimisation but more so by the 'symbolic' aspects of a given policy. This, he maintains, is particularly true for crime policy, whereby 'symbolism' is employed/manipulated for the purposes of political mobilisation. If we consider New Labour's approach to crime during this period, there was perhaps no slogan more symbolic than Blair's 'tough on crime, tough on the causes of crime'.

The phrase 'tough on crime, tough on the causes of crime' is undoubtedly one of the most famous political slogans of the past few decades (Green et al., 2005). Invented by Tony Blair, and spoken for the first time in a radio interview in 1993 (Burney, 2005: 18), the phrase captured a political and social zeitgeist, satisfying not only those who were advocating greater personal responsibility and 'stronger' accountability for criminals, but also those who identified the problems associated with social stratification, poverty and disadvantage. In one fell swoop, Tony Blair had apparently appealed to New Right sensibilities as well as to the concerns of social and psychological determinists.

However, in 1993, in the wake of the abduction and brutal murder of the toddler Jamie Bulger by two 10 year old boys, Robert Thompson and Jon Venables, the country had become increasingly concerned about a society which was seen as becoming 'lawless'. In particular, public anxiety became focussed upon youth crime which was viewed as 'spiralling out of control'. The infamous grainy CCTV image of Thompson and Venables leading the toddler by the hand from a shopping centre was broadcast around the nation and became symbolic of a broader 'moral panic' about depreciating social values and moral decay.[13] The heightened social climate at this time had, according to Burney (2005: 19) a significant impact upon the direction of Blair's 'tough on crime...' agenda.

Indeed, Burney (2005: 18–19) asserts that it was this atmosphere of anxiety that pushed Blair to 'translate his belief in mutual responsibilities into a "no excuses for crime" narrative. ... His communitarian beliefs might have taken a different turn. ... But instead the culture of censure and blame took over'. Similarly, Downes and Morgan (2007: 214–15) have suggested that the 'tough on crime' slogan adopted by New Labour was in effect rather disingenuous, particularly, they contend, because the phrase served to conceal the de facto practice and policy-based difficulties inherent in such a dual agenda. Indeed, they propose that 'whilst liberal reformers had interpreted "tough on crime" to mean target-hardening and other forms of situational crime prevention, and "tough on the causes of crime" to mean tackling poverty, unemployment, and inequality, in reality it largely meant being tough on the *criminal'* (p. 215, original emphasis). Citing New Labour's subsequent introduction of a litany of 'largely punitive measures' – including the ASBO – Downes and Morgan conclude that the effect was to dissipate any possible criticism that New Labour was 'soft on crime' (p. 215).

And so it was, that when New Labour did ascend to office in the 1997 general election, the Party continued their trend of hegemony on law and order, enacting new legislation (most notably, the Crime and Disorder Act 1998) and invoking mandatory sentencing (as described above), alongside the introduction of a range of other 'punitive' measures designed as part of the broader 'tough on crime' agenda (see, for example, Morgan and Newburn (2007) on the development of youth justice). In particular, New Labour maintained a strong and unwavering focus upon community disorder, in the guise of 'anti-social behaviour'. Continuing to propagate the notion that rights equate to responsibilities, underpinned by principles derivative of the 'Broken Windows' thesis, a succession of Labour Home Secretaries promoted the discourse of community shared values on acceptable behaviour in publications such as *No More Excuses* (1997). In the government's White Paper, *Respect and Responsibility – Taking a Stand Against Antisocial Behaviour* (2003a: 4), the Home Secretary David Blunkett argued – in a clear reflection of Tony Blair's earlier statements on community disorder – that:

> As a society, our rights as individuals are based on the sense of responsibility we have towards others and to our families and communities. This means respecting each other's property, respecting the streets and public places we share and respecting our neighbours' right to live free from harassment and distress. It is the foundation of a civic society…. [A]nti-social behaviour, in whatever guise, is not acceptable and…together we will take responsibility to stamp it out, whenever we come across it. This responsibility starts in the family, where parents are accountable for the actions of their children and set the standards they are to live by. It extends to neighbours, who should not have to endure noise nuisance. It continues into local communities, where people take pride in the appearance of estates and do not tolerate vandalism, litter or yobbish behaviour. Our aim is a 'something for something' society where we treat one another with respect and where we all share responsibility for taking a stand against what is unacceptable. But some people and some families undermine this. The anti-social behaviour of a few, damages the

lives of many. We should never underestimate its impact. We have seen the way communities spiral downwards once windows get broken and are not fixed, graffiti spreads and stays there, cars are left abandoned, streets get grimier and dirtier, youths hang around street corners intimidating the elderly. The result: crime increases, fear goes up and people feel trapped.

New Labour's campaign to tackle anti-social behaviour – but also Tony Blair's wider initiative to end the so-called '1960s liberal consensus on law and order' – has been criticised by many academics as opportunistic policy-making, influenced by the politicisation of crime and criminal justice. Indeed, a number of commentators have situated New Labour's approach to law and order very firmly within the theory of popular punitivism (see, e.g., Garland, 2001). Understood as embodying attempts to sate public desire for increased 'toughness' in law and order policies (underpinned by media driven (mis)characterisations of crime and its prevalence), New Labour policy on anti-social behaviour is often identified as being driven by popular public sensibilities and demands for increasingly more punitive action, rather than as a genuine, empirically grounded policy initiative (Cavadino and Dignan, 2002; Morgan and Newburn, 2007; Raynor, 2007).

The popular perception that law and order policy has become much more 'punitive' and 'authoritarian' is in part explained by critics' accounts of New Labour's political (re)positioning and (re)branding. As Downes and Morgan (2007: 210) contend: 'Labour has arguably morphed into a party of the Centre Right rather than the Centre Left, and is implementing "Thatcherism" by other means – a cooler, managerialist version ... New Labour have embraced policies and principles associated with neo-liberal rather than social democracy, of the Right rather than the Left'. So does this then mean that anti-social behaviour policy is underpinned by principles of the Right? How should we locate anti-social behaviour policy politically and ideologically? In the context of the political landscape of anti-social behaviour policy, it is instructive for us now to consider how the policy might best be characterised.

Anti-social behaviour and the politics of 'Left' and 'Right'

As we have already seen, New Labour's 'tough on crime' agenda was bifurcated in its objectives – appealing to both traditional social democratic principles and New Right sensibilities. However, with the appointment of a new Prime Minister (Gordon Brown), and more recently, the appointment of a new Home Secretary (Alan Johnson), these changes undoubtedly have consequences for the development and evolution of anti-social behaviour policy. While commentators had contemplated that a change in the Labour Party leadership might signal a more nuanced and less authoritarian policy perspective on anti-social behaviour and, in particular youth crime, this has manifestly not transpired. As we have seen, the incoming Home Secretary announced a new 'drive' to tackle anti-social behaviour and has set out plans for the introduction of measures to speed up the ASBO process, to make it more accountable to local people, and to ensure improved support networks for victims. In this context, critical commentators will likely view the new government 'drive' on anti-social behaviour as part of New Labour's neo-liberal 'Right-leaning' or 'Centre-Right' law and order policy continuum.

And here is where the difficulty lies. To term New Labour's anti-social behaviour policy as 'Centre Right' or indeed, as embodying those 'principles ... of the Right' (Downes and Morgan, 2007: 210), is a popular argument within critical criminology. In reading critical scholarship which makes these categorisations of anti-social behaviour policy (and law and order policies more generally), one might reasonably reach the conclusion that 'Right', 'Centre Right' and 'Right-leaning' are necessarily being used in a pejorative sense, particularly in those works where anti-social behaviour and the Respect Agenda are defined as 'a punitive, exclusionary approach which bears down most heavily on the already stigmatised and marginalised groups in society' (ibid: p. 220). Two questions then arise: what do we understand by the terms 'Right leaning', and 'Centre Right' and, is it in fact accurate to identify New Labour's approach to law and order and, more specifically, its anti-social behaviour policy, as underpinned by 'principles of the ... Right rather than the Left' (Downes and Morgan, 2007: 10).

There has undoubtedly been a tendency within criminology, to erroneously contrast conservative and liberal-thinking (and their approaches to criminal justice policy) as if each were the antithesis of the other. A number of scholars have sought to establish that, historically, liberal (associated with the 'Left') and conservative (associated with the 'Right') ideologies exist at two polar opposites on a continuum scale (Rosch, 1985; Walker, 1985). For example, Wilson and Ashton (1998: 8) distinguish the difference between 'left' and 'right' criminological theories as follows:

> Right-wing theories tend to blame human wickedness and greed, permissive social policies, sexual freedom, the media, family breakdown

and lack of respect for authority. In contrast, left-wing theories emphasise the role of social and economic factors, materialism and lack of support.

Similarly, they characterise the differences between *political philosophies* equally succinctly:

> Conservatism is an individualist doctrine which holds the individual to be free and rational, and therefore entirely responsible for his or her own actions. The parties of the left and centre, by contrast, believe that individual actions are shaped not only by individual will, but also by the broader social and economic context in which they occur. (p. 11)

Thus, within this paradigm, liberal-thinking is identified as orientated towards rehabilitation and is increasingly cynical about developments in criminal justice, while conservative-thinking is characterised as an intractable 'get tough' approach to law and order (see e.g., Walker, 1998). Braswell and Whitehead (1999: 50) observe that indeed the popular view is often that: 'conservatives seem to increasingly support greater police powers and more arrests while liberals appear to be more supportive of current community-based policing that encourages intervention in family crisis situations and empowerment of communities'. Yet such an account of liberal/conservative thinking is not only overly simplistic but is also inadequate as a framework within which to understand the subtleties and complexities of developments in criminal justice policy and practice.

Kania (1988: 74) has gone further in arguing that the use of the word 'conservative' as 'a pejorative by those who are not...clouds the word and the complex ideas it represents with a negative imagery'. Asserting the existence of a historical (sociological) tradition of associating conservatism with draconian and disproportionately punitive criminal justice policies, he cites Kinloch's (1981) typology of ideologies in sociology as illustrative of a broader misrepresentative and overtly negative characterisation of conservative or 'Right-leaning' thinking. Indeed, Kania observes that Kinloch uses terms such as the 'reimposition of community' and 'authoritarian political solutions' to 'reimpose social order' in his description of conservatism while describing liberalism as inculcating 'reason', 'logical analysis', 'individual freedom' and 'pragmatism' (Kania, 1988: 75, citing Kinloch, 1981: 28). Rather pessimistically, Kania (1988: 76) concludes that 'as long as conservatism remains defined by those whose goal it is to defile it, its defence is most difficult. This has been especially true within criminal justice.'

However, as we have seen, over the past three decades, conservative and liberal approaches to law and order increasingly began to overlap – both in terms of their desire for 'toughness' but also in terms of their perspectives on the 'social' character of crime. Spelman (1994: 312) has argued that crime 'can never be substantially reduced through incapacitation alone'. Citing the importance of deterrence and rehabilitation, and the impact of poverty and unemployment, abuse and marginalisation, Spelman observes the value of incapacitation but also the influence of wider socio-economic policies in ameliorating the continuing difficulties resulting from social stratification and disadvantage.

Moreover, in a recent review of the effectiveness of liberal ('social support') and conservative ('social control') policies on crime in the US during the 1990s, Ren et al., (2008) found that although the two perspectives did exist on a single continuum – with conservatives advocating greater social control and deterrence at one end and liberals calling for increased social support interventions to tackle poverty, unemployment and inequality at the other end – the findings on which measures (social control versus social protection) were effective in reducing crime, overlapped. Conservative and liberal policies cannot now be so easily contrasted. Both conservative and liberal policies on crime identify socio-economic explanations of criminality. As a result, Green et al. (2005: 7) rightly observe that 'the division is not between those who emphasise underlying social causes and those who do not; it is more a debate about which underlying social causes are important.' Nonetheless, the view persists that the New Labour approach to law and order is in some way underpinned by principles originating from the Right but to what extent is this a fair explanation of New Labour's policy approach?

It has been suggested that the old Labour Party's shift to New Labour, and its adoption of Third Way politics, signalled an attempt to move beyond the somewhat archaic definitions of 'Left' and 'Right' towards a new political philosophy that is embedded in a form of benevolent pragmatism which simply asks whether each policy works in practice. Giddens's conceptualisation of the Third Way necessarily rejects the boundary claims of traditional 'Left', 'Right' and 'Centre' philosophies and argues that solutions to problems should emerge pragmatically, spontaneously – without being constrained by historically predetermined political categorisations (Giddens, 1994; 1998). According to Giddens (1998), the Third Way is a response to global change – in particular, to social and economic changes, and the process of globalisation. Far from being rooted simply in political opportunism, the ideology of the Third Way should, as far as Giddens is concerned, be seen as a necessary, deliberate, reflective and rational response to global socio-economic changes. As such, twenty-first century Britain is a complex and multifarious system of modernity where policy and practice cannot be delineated simply into those pertaining to 'Left' and 'Right': policy decision-making is intended to be ostensibly borne out of pragmatic utilitarianism. In this vein, Reiner (2000: 85) observes that New Labour policy on law and order

> ...is novel and cannot be dismissed as either simply punitive or liberal. It is based on an intelligence-led, problem-solving approach, with systematic analysis and reflexive monitoring built into policy development. Its intellectual basis is a thorough review of the evidence concerning the effectiveness, costs and benefits of the main strategies for dealing with offending: preventing the development of criminality by early childhood intervention and education, situational and community crime prevention, policing, sentencing and alternative penal techniques.

Hence the politics of the Third Way is underpinned by a 'what works' strategy, which has been described as 'managerialist' in the sense that managerialism espouses the notion that necessarily competing values can be replaced by pragmatic policy-making. As I have argued elsewhere, anti-social behaviour management exists as risk assessment, and is a form of institutional control that is fragmented; externally constrained; and managerial (Donoghue, 2008). Anti-social behaviour policy, in the same way as other New Labour policies (for example, youth justice policy) is fundamentally an *actuarial* policy, concerned with the quantification and management of risk to achieve 'what works'. These processes of identification, quantification, planning and protection have engendered a system in which performance, accountability and transparency become key elements of regulation (see e.g., Power, 1997).

However, the prevalence of 'risk management' in New Labour policies through the 'expert' identification of 'risk categories' (Kemshall and MacGuire, 2001; Kemshall, 1997) has been criticised for putting 'the lessening of risk, not the meeting of need' in a fundamental position of primacy (Culpitt, 1999: 35). In this sense, technical and professional knowledge(s) are paramount in defining, identifying, quantifying and managing specified 'risk categories' (Scourfield and Welsh, 2003: 404).

Similarly, Reiner (2000: 1) sees Labour's Third Way perspective as inadequate as a framework within which to effectively address social and economic inequality. Railing against the actuarial foundations of New Labour's policies, he posits that:

> Current 'third way' policies for crime reduction may achieve modest success, in part because they indirectly encourage agencies to manipulate statistically recorded outcomes to their advantage. They do not however tackle the underlying sources of crime in the political economy and culture of global capitalism, offering only actuarial analyses of risk variation, and pragmatic preventive interventions to reduce these. In the absence of any broader changes to the social patterns which generate high-crime societies the prospect is of marginal palliatives for crime, which themselves have the dysfunctional consequences of increasing segregation, distrust and anxiety.

Hence the 'what works' approach – which is identified as a derivative of the Third Way political agenda – has been criticised for being too concerned with 'the now', for prioritising policy initiatives that are only worthwhile in the short term and, for being disinterested in pursuing policies that are intended to address social stratification and socio-economic inequality in the long term (Reiner, 2000). Critics argue that the New Labour approach to law and order is thus essentially neo-liberal in the sense that it is a punitive and authoritarian policy that has very limited interest in tackling social exclusion and poverty.

For the purposes of my argument in this book, I do not characterise New Labour policy on law and order in this way. New Labour has not been *disinterested* in social regeneration. For example, New Labour has supported increased investment in child welfare programmes, education, housing, health and youth services to help reduce crime and anti-social behaviour. It is also worth noting that prior to his appointment as Home Secretary, Alan Johnson appeared to signal his ambition to further inculcate *traditional* (old) Labour values on social justice and equality into New Labour philosophies on law and order. In a speech in 2006, the argued that 'our hatred and determination to eradicate poverty is the glue which holds us together: connecting our past with our future; our ideological and pragmatic wings; distilling Old Labour and New Labour into Real Labour'. But how far New Labour has been *successful* in tackling inequality and disadvantage – and how one chooses to measure this – is an entirely separate and distinct issue. The real area of controversy with regard to anti-social behaviour policy, I would suggest, does not necessarily concern whether or to what extent New Labour is interested in or has been successful at addressing socio-economic disadvantage – the real area of controversy concerns *personal responsibility* for anti-social conduct. There are few that would deny that it should rightly be a concern of modern government to address poverty and disadvantage. However, what is most contentious, specifically in the context of anti-social behaviour policy, is whether programmes of neighbourhood renewal should be seen as de facto *precursors* to tackling anti-social behaviour. Hence in the next two chapters we will consider to what extent the notion of civic obligation – the duty to behave reasonably to one's neighbours – should be viewed as conditional upon social renewal and efforts to address social exclusion and disadvantage. First, however, we must consider in more detail the historical background to anti-social behaviour policy, and to what extent modern preoccupations with disorder can be understood in the context of neo-conservative concerns about the 'banishment of outsiders' which necessarily reflect class differences.

4
Anti-Social Behaviour: The Historical Landscape

Given the prominence of the term 'anti-social behaviour' in contemporary social, political and academic discourse(s), one could be forgiven the assumption that anti-social behaviour is solely a concern of contemporary Britain. However, anti-social behaviour exists as a perennial and recurring feature of social life (Donoghue, 2008). A substantial body of scholarly work locates urban disorder, incivility, and criminal and sub-criminal nuisance behaviours as salient issues of social concern for previous generations (Pearson, 1983; Shaw, 1931; Whitehead, 2004). Yet, in terms of aetiology, anti-social behaviour is far from unambiguous (Squires, 2006: 158) since 'anti-social behaviour' is a generic term, and does not necessarily relate to criminal or sub-criminal behaviour – 'anti-social' can of course mean one who is introverted and misanthropic and/or one who is either without, or in possession of poor, social instincts.

If we apply the modern, elastic interpretation of the term 'anti-social' to eighteenth and nineteenth century Britain, for example, it would seem to suggest that 'anti-social behaviour' (in all its diverse embodiments) was ubiquitous and pervasive (Donoghue, 2008). In urban areas, there was a proliferation of criminal gangs, prostitutes and pickpockets while smugglers, rustlers, tavern brawlers, and itinerant beggars were to be found in the rural dwellings. Also becoming apparent in the nineteenth century were the antecedents of new forms of 'anti-social behaviour' such as grave robbers, child pickpockets, and 'hooligans' (Pearson, 1983). Indeed, it is difficult to find a period in at least the last 150 years where there is not historical evidence of social/moral preoccupations and concerns about forms of disorder that we might now classify as 'anti-social'.

By way of illustration, if we consider that seemingly most contemporary of concerns associated with anti-social behaviour: that is, binge-drinking and alcohol- related disorder, we can see how these same concerns manifested themselves in eighteenth and nineteenth century Britain, particularly with regard to the 'gin-craze' of the early eighteenth century. Eighteenth century Britain was essentially a nation of binge drinkers and alcoholics, with substantial levels of alcohol-related violence. Gin had been introduced into Britain from the Netherlands during the 1690s and became so popular that by 1751, Londoners were drinking eight million gallons a year – which works out at roughly a gallon per person. Bernard Mendeville observed that the epidemic of gin drinking was primarily a result of the drink's cheapness. He noted:

> Gin is sold very cheap, so that People may get muddled with it for three half pence and for three pence made quite Drunk even to Madness. ... [so that] it comes within the power of Common People to purchase the hopeful Reputation of getting drunk at a very small Expense.

Although gin-related intoxication was typically associated with the poor, drunkenness was in fact widespread among all social classes at this time. However, eighteenth century concerns about drunkenness centred on the excessive alcohol consumption of the *poor*, rather than the affluent, 'genteel' classes. An illustration of the stratification of anxiety about drunkenness – premised upon class distinctions – is provided by two of the Georgian artist William Hogarth's most famous works: *Beer Street* (1750) and *Gin Lane* (1751). In *Beer Street*, the consumption of alcohol by artisans is portrayed as restrained and enlightened – a civilised pleasure of the professional classes. However, in *Gin Lane*, by contrast, desire has run insane, having become a form of spurious pleasure for the poor. Indeed, Hogarth's *Gin Lane* reflects 'moral panics' of this time about the mania and the moral debauch associated with the gin-craze, which was bolstered by the existence of documented cases about profligate individuals. For example, a notable case of this time is the indictment of Judith Defour in 1734, who was convicted of murdering her infant daughter Mary by choking and strangling her with a handkerchief so that she could sell the infant's clothes for a shilling and four pence to buy gin. As the Chaplain of Newgate observed following her execution: 'She drank and swore much, and was averse to Virtue and Sobriety, delighting in the vilest Companies, and ready to Practice the worst of Actions'.

In a clear parallel with our contemporary anxieties about binge-drinking, the drinking habits of the poor were characterised as a social evil, while the drunken antics of the wealthy were seen as entertaining, merry and light-hearted. Attitudes to drunkenness were thus a reflection of class differences. (It is worth considering that it is only in the last couple of years that 'middle-class' drinking has become identified in policy discourse as a 'social problem'. In a report for the Department of Health (2007) it was found that that people living in relatively affluent areas are more likely to consume dangerous levels of alcohol than those living in deprived areas. Some of the country's most wealthy areas were found to have the highest numbers of 'hazardous drinkers'. Runnymede in Surrey and Harrogate in North Yorkshire came top of the league tables for local authority areas with the highest levels of hazardous drinking. Excessive alcohol consumption has since been labelled the new 'middle-class vice'.)

So while binge-drinking, public intoxication, and alcohol-related disorder provides an example of one historical form of 'anti-social behaviour', there is also much historical evidence documenting that public disorder, hooliganism, gang violence, rioting, youth crime and delinquency (particularly with regard to young male apprentices) were also specific public concerns during the eighteenth and nineteenth centuries. Hence, 'anti-social behaviour', as an embodiment of disorder and incivility, is in many ways not a 'new' phenomenon. And interestingly, we can see that many of the contemporary concerns surrounding anti-social behaviour regarding parenting, education, unemployment, social breakdown, the cost of alcohol, 'feral youths', the moral fortitude (or otherwise) of 'the poor', are also reflected in historical discourses about urban disorder. In this case, why has anti-social behaviour become so hugely elevated in the public consciousness in recent years? Is this attributable solely to a protracted and robust political and media campaign which has sought to highlight anti-social behaviour and its consequences as a quintessentially modern social phenomenon which is 'spiralling out of control'? Or is the current preoccupation with anti-social behaviour representative of a genuine substantive *increase* in urban disorder and problem behaviour?

The purification of space

While disorder and nuisance have necessarily featured as historical social milieu, the twentieth and early twenty-first century appears to have accommodated a conversion of sorts: whereby public concerns about anti-social behaviour and moral decay (for the two are often inherently bound together in media and public discourses[1]) transgressed 'acts' to become about forms of culture (Donoghue, 2008). Contemporary concerns about urban disorder and incivility are now paired with a more aberrant cultural anxiety, centring on forms of *community* related troubles (ibid.). Moreover, political discourse identifies the need to protect 'communities' from anti-social behaviour. As such, the concept of 'community' is one that is central to anti-social behaviour policy – reflecting modern constructions of crime control problems and solutions defined through the use of the terms 'community' and 'communities' within social policy.

Jock Young's (2001) critique of the modern political understanding of the term 'community' (and its constituent elements), centres on its potential for use as a tool of social control when the terminology is embedded within criminal justice policy. Consequently, it has been argued that the tendency has emerged for a unilateral, cross-party acceptance of how communities are to be defined, based upon reductive, depoliticised terminology which is limited in its ability to be inclusive of the tenuous, rarefied subtleties and often deeply recessed variations in local contexts (Stenson and Edwards, 2004). Moreover, Edwards and Hughes (2005) have posited that the current (and unique) behavioural focus placed upon both youth disorder, and measures to prevent persistent offending within 'communities', has subsequently eclipsed dominant social democratic interventions in local structures of social inequality and exclusion that had previously been pursued by municipal local authorities, in part reflecting Young's original contention that the term 'community' has become a means for domination, stratification and an instrument for social exclusion (2001).

However, the identification of the 'community' as a specific locus of governance has also resulted in what has been termed a 'new localism' whereby neighbourhoods become the sites of policy intervention underpinned by the promotion of community engagement and active citizenship (Forrest and Kearns, 2001). As Flint (2006: 26) observes:

> The rise of neighbourhood governance combines the concept of community with a long-standing emphasis on area-based regeneration to tackle the spatial concentration of localised social problems.... One important element of the transfer discourse is the facilitation of tenant participation in the management of their homes and local communities....

Hence, it has also been argued that the emphasis placed upon community governance and engagement with anti-social behaviour policy interventions has resulted in the empowerment of communities rather than, alternatively, an oppressive and encroaching process of 'responsibilisation' (Garland, 2001). Without doubt, however, as Steventon (2006) amongst other commentators have argued, the emergence of anti-social behaviour as a central strand in political discourse has facilitated increasing political intervention in community issues.

Drawing upon Cohen's (1985) analysis of the creeping, rhizomatic development of social control and its changing character, it has also been suggested that within the contemporary anti-social behaviour discourse, there has begun a shift in focus from preventing, or at least attempting to improve social conditions which had been identified as fostering crime and sub-criminal behaviour to, conversely, a redefined pursuit of and fixation with 'the management of criminal behaviour sequences' (Bannister et al., 2006: 925). In this way, critics have argued that the policy focus in recent years has subsequently moved from being concerned with security and the improvement of social conditions/social capital (that is, the removal of the holistic elements that were present) to a process of social control concerned, principally, with 'the purification of space both commercially and residentially' (Bannister et al., 2006: 925). This programme of gerundive urban 'sanitising', it has been suggested, is moulded around socially subversive constructions of difference – or 'otherness'. In this way, the contemporary preoccupation with the sequestration of behaviours that are categorised as deviant or indeed, normatively suspect, can be understood, it has been proposed, in the context of the 'risk society' thesis (Beck, 1992; Giddens, 1990).

Anti-social behaviour as 'risk'

The identification, classification and regulation of risk is now very much a modern fixation which permeates many different facets of our daily lives. From terrorism to natural disasters to allergies to the stock market, risk and how it might best be 'managed' is a theme which is pervasive in modern life (Donoghue, 2008). Ekberg (2007: 343) contends that the risk society thesis has succeeded in achieving 'communities united by an increasing vulnerability to risk... The risk society expands the traditional concept of risk understood as the sum of the probability of an adverse event and the magnitude of the consequences, to include the subjective perception of risk, the inter-subjective communication of risk and the social experience of living in a risk environment... it is not just health and the environment that are at risk, but in addition, the fundamental socio-political values of liberty, equality, justice, rights and democracy are now at risk'. Hence, the *hazards* resulting from life in modern society are now dealt with through a systematic process of risk quantification and management (Beck, 1992).

And indeed, the contemporary approach to anti-social behaviour management embodied within government policy reflects an overarching social governance approach to anti-social behaviour that is concerned with 'risk' management. The identification of risk-bearing behaviours is central to processes of anti-social behaviour control. However, forms of risk prediction have been criticised for being de facto processes of 'victim blaming' (Griffin, 1993). Armstrong (2004) further developed those existing critiques by proposing that the dominance of 'risk' had resulted in the growth of crime prevention policies that are fundamentally concerned with the construction of notions of deviance, the criminological 'other', and early intervention policies centring on the (re) socialisation of parents and their children in line with normative (and generally middle-class) standards of acceptable behaviour.

In a similar vein, Edwards and Hughes (2005: 347), for example, argue that 'contemporary political and social concerns about "public safety" – while illustrating dominant social democratic interests in harm and victimisation – also exemplify neo-conservative concerns about the '"banishment" of outsiders'. It is the format, construction and problematisation of, in wider terms, 'crime' and 'security', that Edwards and Hughes contend promotes both strategies of control (in this case, the purification of space) that subsequently, or necessarily, obviates particular social groups. As Vaughan (2000) has identified, it is the economically marginal who are most often perceived as 'other' – an observation that is particularly relevant to the 'urban renaissance' (and subsequently, the anti-social behaviour) paradigm. For example, in a consideration of Williams' conceptualisation of urban culture (2004), Bannister et al. (2006) detail the nature of Williams' analysis of the 'improvements' being made to city centres – which, it is suggested, are concerned with 'a predominantly bourgeois culture involving consumption by the privileged and leisured' (p. 923).

Similarly, Sennett (1996) has identified what he has termed the 'myth of the purified community', whereby individuals *sharing similar social and economic characteristics* have increasingly begun to seek solace in traditional notions of the 'community'. These communities are intended as vestiges of 'ontological security' (Giddens, 1991; 1992), protected from urban disorder, but are also, according to Sennett, the loci of increasing intolerance towards difference. Hence, critics have maintained that the pursuit of new 'purified' communities and better cities (particularly town centres), has moved from being concerned with achieving security, urban regeneration and increasing social cohesion, to encapsulating punitive forms of 'cosmopolitan intolerance' – centred on the removal of difference and 'otherness'. Peoples' sense of socio-economic insecurity is thus identified as fostering desires for increasingly punitive action on urban disorder (see, for example, research conducted by Maruna and King, 2004).

While the sceptical disposition of many academic critiques of anti-social behaviour management towards state strategies of security and community safety are, as we have seen, underpinned in large part by an overarching concern with security as 'social control', it is also important to understand strategies of risk management in the context of their potential to create ontological security. Giddens's (1991) notion of ontological security identifies the primacy of individuals in giving meaning to their lives: people are able to give meaning to their lives through the experience of continuity and order, which are reproductive of positive emotions determining a sense of calm and freedom from anxiety. To put it simply, where individuals are able to experience stability in their lives, and freedom from chaos and danger, then they will derive ontological security. Moreover, ontological security is about reasonable expectation of future events: an individual's ability to have a positive view of the world, and future happenings, contributes to their sense of ontological security. If one feels fairly certain about what will happen in the future, one feels more secure. Instinctively, one is likely to feel insecure if one cannot predict within any degree of certainty what is likely to happen from one day to the next.

While the notion of ontological security is complex and multifarious, Loader and Walker (2007: 166) provide an instructive account of those essential elements which must be present in order for ontological security to manifest itself:

> A sense of dignity and authenticity, of ease with and acceptance within one's social environment, are crucial to ontological security: and, as we have seen, it is these very aspects of social identity that are implicated in the collective self-understanding attendant upon the accomplishment of political community. In other words, to be a member of a stable political community and to feel oneself confident in that sense of membership is already to raise one's threshold of vulnerability – to possess crucial resources in the management of fear and avoidance of security anxiety.

In this sense, security, *and interventions to enable security*, might also be understood as a 'thick public good' (Loader and Walker, 2007) and not simply as a product of (negative) social control. As such, the notion of ontological security should not be understood as relating only to those modern grand narratives pertaining to global security, but it might also be examined in the context of local narratives – protecting individuals, in their day to day lives, from fear, stress and anxiety resulting from anti-social behaviour and community disorder. Indeed, Webb (2006: 45) advocates that 'the logic of security should be stretched to include safety, vulnerability, coping strategies, social support and protection'. Similarly, Ewald's (1991) analysis of security and the welfare state as 'providential' goes beyond linear analyses which link expert interventions unilaterally with social control. Indeed, Ewald sees something positive in the role which the State can play in ensuring the provision of safety and the freedom from anxiety. Correspondingly, Beck (1999: 226, original emphasis) observes that:

> Ewald's theory marks a significant shift in the interpretation of the welfare state. While the majority of social scientists have sought to explain the origins and construction of the welfare state in terms of class interests, the maintenance of social order or the enhancement of national productivity and military power, Ewald's argument underlines the provision of services, the creation of insurance schemes as well as the regulation of the economy and the environment in terms of the *creation of security* ... experts play a central role in answering the question 'how safe is safe enough?'.

In this sense, expert interventions in the domain of social welfare are not directly equated with social control (and its parallel connotations of domination, stratification and the deprivation of liberty). Instead, Ewald's analysis sees social welfare interventions as *potentially positive* responses to events/circumstances in contemporary society. In this way, as Webb (2006: 43) notes, these emerging expert welfare systems can be 'both limiting and enabling'. Importantly, such a view does not determine that state interventions concerned with increasing security are de facto positive, rather these analyses contend that expert mediation may potentially expound negative ('limiting') *or* positive ('enabling') consequences for individuals/groups, or indeed, a mixture of both. Intervention is not, however, seen as necessarily or expressly connoting negative social control.

Theoretical perspectives on anti-social behaviour management are often predisposed to annotate anti-social behaviour as a 'resource' within the risk society paradigm – that is, as an 'auditable' (Parton, 1998) agent proxy whereby 'risk bearing' behaviour(s) is coercively sequestered in order to increase security resulting in the marginalisation and exclusion of disadvantaged populations (Donoghue, 2008). However, while the emphasis is upon anti-social behaviour as a prohibitive, prescriptive vehicle for social control; what is overlooked by these analyses is the *ascendancy of the individual*. Structuralist accounts of the use of ASBOs, and more widely, anti-social behaviour policy, tend to understate the importance of individual agency in contributing to policy administration and outcomes. For example, community responses to (perceptions about) the existence of anti-social behaviour have resulted in individuals becoming engaged in social processes to reduce 'risk bearing' behaviour(s). As a result, anti-social behaviour community action groups, youth centres, community clean-ups, residents groups, neighbourhood watch schemes and community organised diversionary activities are to be found nationwide.

While the few evaluations of these processes that are in existence have been carried out locally and with little standardisation in methodology (Armitage, 2002), and limited empirical research evidence exists on the quantifiable effectiveness of ASBOs themselves (NAO, 2006), individuals and groups have anecdotally reported greater social cohesion, and a sense of 'empowerment' in contributing to community safety (Home Office, 2007). For instance, residents and groups are involved in information sharing with the police and local authorities; in evidence-gathering (for the purposes of ASBO applications in their communities); and in encouraging other residents/individuals to become engaged in community-based anti-social behaviour initiatives (Donoghue, 2008). In this respect, individuals are reflexively making themselves 'the subjects of social processes' (Ferguson, 1997) that are regulating anti-social behaviour.

Moreover, Casey and Flint's (2008: 113–14) research found that the importance of the community's 'voice' in identifying their experiences and expectations is pivotal:

> [There is] evidence that communities in neighbourhoods perceived to be 'at risk' need the support of agencies such as the police in developing a collective efficacy to address anti-social behaviour. Key to this is the idea of 'voice', which is defined as the ability and capacity to articulate a community's self-defined needs, and to have those

listened to and taken seriously by those in authority (Innes and Jones, 2006: 41).... [I]t is clear that the police and other agencies of social control need to think about how their interventions can have a positive impact on the capacity of local communities to practise formal and informal social control over the longer term. Empowering residents individually through facilitating and supporting individuals to report incidents, and collectively through community mediation techniques and community consultation, as used in 'hot spot' initiatives in some of our study areas, to identify residents' knowledge and priorities, are important mechanisms here.

Moreover, the individual as 'service user' has also become a service evaluator (Lash and Urry, 1994: 4), who participates in 'the critical appraisal of social and policy processes and who possesses heightened expectations and demands of these processes in their quantifiable/demonstrable "effectiveness"' (Donoghue, 2008: 350). In this way, it can be said that the state does not have a 'monopoly on promoting empowerment' (Ferguson, 1997: 231), in the management and control of disorder, or even in defining anti-social behaviour (Donoghue, 2008).

However, as Loader and Walker (2007: 91) identify, structuralist academic perspectives tend to reflect the pervasive state scepticism inherent in sociological writing on security and policing. As such, these accounts display

...a structural fatalism that overlooks the overlap between the production of specific and general order, such that disadvantaged groups and communities have a considerable stake not only in controlling state power, but also in using public resources (including policing resources) as a means of generating more secure forms of economic and social existence.

In a similar way, anti-social behaviour management has been duly criticised on the basis that it forms part of a wider political (and social) agenda that is underpinned by an inversely punitive criminogenic approach to marginalised social groups (the young, the economically disadvantaged, and so on) by politicians and majority social groups in pursuit of urban aesthetics. An alternative view can be progressed however, that anti-social behaviour management might *also* be understood in the context of what role it plays in establishing ontological security for both vulnerable people and the wider community – and not simply in the context of *punitive* social control. Flint (2006: 22), for example, contends that increased tenant participation in anti-social behaviour management should be seen as embodying *positive* aspects of individual empowerment and self-governance:

> This attempt to encourage autonomy and self-regulation is an important dimension within the contemporary politics of behaviour. It involves an understanding of tenant responsibility being framed as a proactive and empowering process within housing management, and the conceptualisation of the responsible and responsive tenant as an increasingly central figure in the organising mechanisms of housing governance... These behaviours include engagement in community activity, volunteering, involvement in the strategic decision making of housing organisations, and a more active role in the governance of anti-social behaviour. ... What we are witnessing... *is the continuing reconfiguration of the identities of social housing tenants from passive and dependent welfare recipients into autonomous, empowered and responsible individuals....* (emphasis added)

As Harrison (2001: 103) contends, academic critiques (of behavioural intervention strategies) which identify such measures as 'authoritarian landlordism' fail to recognise that 'tenants themselves are disempowered by violent, racist or criminal neighbours'. Equally, Ferguson has argued that traditional critiques of risk theory in late-modern social work view 'risk' in wholly negative terms and 'little or no attention is given to conceptualising social interventions or risk... in terms of the new opportunities for constructing a "well protected self"' (Ferguson, 2001: 48).

Nevertheless, there remains a pervasive academic scepticism that 'security' (and social policy interventions aimed at achieving security) necessarily intersects with social stratification and subsequently, notions of exclusion and marginalisation. As Loader and Walker (2007: 82) observe:

> When confronted with the suggestion that security can be conceptualized as a public good it asks: whose security? which public? which good? It stands quizzically aghast at the idea that forms of trust and solidarity can (or indeed should) be fostered between constituencies with such structurally divergent interests. It asks: what is the point of democratizing security if the rules of the political game are stacked in such a way that certain groups find themselves losing time and again?

And these are of course important questions. Of particular salience is the question of how security (in the context of social protection) and efforts to promote and to achieve it are linked to outcomes of exclusion and marginalisation, specifically in respect of minority social groups. Are efforts to (re)establish 'purified communities' which are devoid of otherness, in fact premised upon concerns about disorder originating from a 'moral panic' about civic values and depreciating moral standards? Has a new 'moral panic' about anti-social behaviour led to discrimination against disadvantaged populations?

A rise in incivility or a 'moral panic'?

In his seminal work *Folk Devils and Moral Panics*, Cohen (1972) describes a 'moral panic' as 'a condition, episode, person or group of persons [who] become defined as a threat to societal values and interests'. He suggests that bouts of moral panic occur periodically, prompting concerns that social and moral values underpinning society may be in danger. In this respect, contemporary concerns about anti-social behaviour have been identified as exemplifying a form of 'moral panic' about behavioural standards and incivility in contemporary Britain. That is to say, another component of the critique on anti-social behaviour management (as it is currently manifested in policy and discourse), is that contemporary concerns about incivility and neighbourhood nuisance are premised upon a somewhat romanticised and detached view of the past which in turn identifies anti-social behaviour as a 'new' and distinct problem in modern social life. Contemporary concerns about anti-social behaviour, it has been argued, form part of a pervasive 'moral panic' whereby levels of crime and disorder remain relatively constant but with increased public anxiety about rising incivility.

Before we go on to consider whether contemporary concerns about anti-social behaviour do indeed represent a 'moral panic' of the kind that Cohen describes (see also Hall et al., 1978; Young, 1971), it is important to address the central criticisms made of the moral panic theory. The theory of moral panic has been attacked on the basis of its phraseology – in particular, the use of the term 'panic'. In evaluating the justification for this form of critique, one must consider whether reaction to particular episodes or social conditions can reasonably be described as 'panics'. A panic can be defined as hysteria or confusion and connotes over-reaction and irrationality. However, critics contend that there may be *rational* bases for fear based upon experience. According to Cornwell and Linders (2002: 307, emphasis added) 'whether or not a social object becomes deviantized depends on a complex process of social construction involving active, *not merely reactive*, efforts by social actors (contrary to traditional images of "panic")'. Moreover, the term 'moral' is contentious, largely because of its ambiguity and its relativity (Boethius, 1994; Miller and Kitzinger, 1998).

Jewkes (2004) has formatted a comprehensive analysis and evaluation of the moral panic theory in which she contends that there are fundamental problems with the use of the moral panic theory in modernity which necessitate its revision and its reconstitution in order to address its inherent difficulties as a viable conceptual framework. Alongside problems related to the terminology of the phrase, Jewkes contends that the concept of moral panic has now become so broad in terms of the multitudinous scenarios in which it is, or has been, applied as to render it conceptually almost redundant. There are also empirical difficulties, she posits, borne out of any claim that there exists a direct causal relationship between media discourse and public response(s). The theory has also been criticised for denying the influence of agency (de Young, 2004; Miller and Kitzinger, 1998) and for over simplifying causality (de Young, 2004).

In view of these difficulties, Cornwell and Linders conclude that the notion of moral panic 'is so laden with ontological and methodological difficulties as to render it virtually useless as an analytical guiding light' (2002, 314). Others who have been critical of the theory assert that it is only its revision that is required to ensure its survival as a valid conceptual framework. Indeed, Critcher (2008: 1138) concludes that 'moral panic analysis is better understood as an ideal type: a means of beginning an analysis, not the entire analysis in itself. And for that no better tool has yet been devised'. Bearing in mind these criticisms of the moral panic theory, is it then appropriate to locate contemporary concerns about anti-social behaviour as a new moral panic?

Geoffrey Pearson has argued in his incisive work *Hooligan: a History of Respectable Fears* (1983) that there is historical evidence of a steadfast and unyielding societal fixation on the perceived attrition of moral standards and social discipline. This preoccupation crystallises in the form of a wider, pervasive belief that crime and urban disorder are perennially escalating. Within this paradigm, he contends that the disproportionate representation of young people, and specifically, young males, in criminal statistics is repeatedly 'rediscovered' as an original feature of crime and disorder. Indeed, we can continue to observe the contemporary pervasive concern with the behaviour of young males – and its historical legacy. For example, the focus of the criminal biographies of the seventeenth and eighteenth centuries was frequently upon the illicit and disorderly activities of young male apprentices which, Pearson contends, was representative of a more general, ubiquitous concern about the 'problem' of youths in London. Philip Rawlings (1992) estimates that there were approximately 20,000 young male apprentices by 1700, and that they had a tradition of radicalism and solidarity. As Rawlings suggests, this seems to sit behind the link between young males, disorder and crime.

Pearson identifies that there was a repeated complaint that deficient apprenticeship was a direct cause of what was beginning to become known as 'juvenile delinquency'. Without the structure and discipline imposed by a dogmatic and exacting apprenticeship, young males were likely to become involved in morally fallacious behaviour, if not actual criminality. This view was reflected in a number of 'conduct books' directed at apprentices, which were published during this period. The conduct books portrayed libidinous behaviour like it was an infectious disease – which young males were particularly predisposed to (Pearson, 1983). However, according to the conduct books, young males could be protected against a descent into immorality, incivility and criminality through a properly regulated apprenticeship. Such an apprenticeship prohibited an apprentice (amongst other things) from getting married, from gambling or going to alehouses, and provided that apprentices were bound to ensure that they did not break the Sabbath (ibid.).

Yet very similar concerns associated with discipline, incivility, moral breakdown and the detachment of young males from society continue to propagate today. But are these concerns – in relative terms – experienced as any more prolific and any more 'tragic' than they were in any other particular historical period that one could identify? In fact, as Pearson rightly determines, similar social and moral preoccupations are continually apparent over the last two centuries. As such, Pearson posits that each historical era has understood itself as existing at a point of 'radical discontinuity' with the past. By way of illustration, if we look back at recent modern history, to the 'sex and drugs' culture of the 1960s; to football hooliganism and punk music in the 1970s and 80s; TV/film/video game inspired violence in the 1990s; and latterly; 'chav' culture,[2] those advocating the moral panic thesis would contend that each era has regarded itself as presenting a uniquely *more problematic* context for decreasing moral standards than in other historical periods (Donoghue, 2008).

Many academics and commentators have argued that Pearson's thesis is extremely valuable in contextualising contemporary concerns about rising crime and incivility. For example, Bannister et al. (2006: 919) contend that 'Pearson's observations provide a useful reminder that moral panics around issues of incivility and disorder in British cities are nothing new (although we might imagine them as new through romanticising the past)'. Moreover, the academic literature on anti-social behaviour (almost without exception) locates contemporary concerns about anti-social behaviour within the context of a historical continuity of urban disorder in Britain. However, Dennis and Erdos (2005) provide one of the few comprehensive rejections of Pearson's work. The primary basis for their critique is Pearson's supposition in *Hooligan* that the Edwardian and inter-war periods were *at least as violent* if not more so than the late twentieth century – which Dennis and Erdos contend is manifestly incorrect. Accordingly, they argue that 'Contrary to what Pearson and other exponents of the "moral panic" theory say, there has been clearly a situation of low crime in the earlier period and high crime in the later period' (Dennis and Erdos, 2005: 52–6).

Dennis and Erdos's view on moral panic, crime and disorder is not a popular one within academic analyses. While their work has been criticised for being empirically unsubstantiated and 'moralistic', it is also likely to be the case that their work attracts such sustained criticism because it attacks the dominant liberal view on crime, and its causes (in particular, Dennis and Erdos's view on the disintegration of the family and its impact upon crime and disorder is seen as especially polemical and 'outdated'). According to Green, Dennis' rejection of the moral panic thesis on crime and disorder 'calls attention to the role of a new class of "conformist intellectuals" in undermining what common sense tells us about rising crime'. As such, Green proposes that the problem with the moral panic view is that it is propagated by 'social affairs intellectuals [who] are strongly inclined to subscribe to the politically-correct doctrines of the day. The result is that universities, instead of being havens for fearless seekers after truth, have become easy berths for conformists who are reluctant to allow the inconvenient facts to spoil a good theory' (in Dennis, 1993: vi).

Dennis (1993; 2004) had also previously set out his position on the prevalence of crime and disorder, and the moral panic theory, and had reached the conclusion that those arguments that asserted that the fear of crime was exaggerated were statistically inaccurate. With the use of official Home Office statistics, Dennis concluded that public perceptions about neighbourhood disorder (including vandalism, graffiti, littering, drug use and drug dealing) had increased between 1992 and 2002, and that evidence from the British Crime Survey (BCS) demonstrated that people living in disadvantaged areas were more likely to report higher levels of disorder than those living in affluent areas. In a blanket rejection of the moral panic thesis, Dennis (2004) argues that by invoking the notion that urban disorder is a form of moral panic, disorderly neighbourhoods continue to be subjected to a process of social decline and degeneration through government/state inaction. In this way, individuals have been: 'Lulled into complacency by assurances that their fears were exaggerated and that crime and disorder were at historically low levels, when they were not mocked for being in an irrational state of moral panic, other areas were left to continue their slow cultural disintegration'.

Arguing for a causal link between perception of crime and disorder, and the *existence* of crime and disorder, Dennis (2004) suggested that if perceptions about levels of disorder

> ...were largely unrelated to reality, and are the creatures of exaggerated fears and moral panic, it would be difficult to explain why respondents in affluent areas, who would be expected to be the more sensitive to a given level of disorder in their immediate localities, in fact reported much less disorder than respondents in poor areas, who might be expected to less sensitive to the same level of disorder. Council estates were perceived by their residents as much more disorderly than were non-council residential areas by their residents.

Although the problems associated with the use of official statistics have already been identified (see Chapter 3), and so the value of official crime data should be considered in the context of its potential limitations, it is important to note that Dennis' argument that higher *perceptions* of disorder in disadvantaged neighbourhoods correspond to a substantive higher *incidence* of disorder, is supported by other empirical evidence which has found that anti-social behaviour is most prevalent in areas of concentrated multiple deprivation (Millie et al., 2005) (the disproportionate incidence of anti-social behaviour in areas of social deprivation and in inner cities will be considered more fully in the next chapter).

Whether one attests to Dennis and Erdos's view that crime (including anti-social behaviour) has increased over the years, or whether one is more inclined towards the supposition that we inhabit a society which is no more adversely affected by anti-social behaviour and urban disorder than in previous centuries, the value of Pearson's theory remains eminent. Above all, Pearson is advocating an abandonment of the apocryphal notion of the *novelty* of street crime and urban disorder. His theory is not, however, dismissive of the legitimate anxieties that surround these behaviours. Pearson contends (writing in 1983) that, 'within the present climate of opinion, if the worries of the present day are at all diminished by knowing that they are in no way unparalleled – then the inevitable complaint will be that one is not taking the problem seriously enough'. However, as Pearson argues in the final part of the book, if the notion of 'novelty' is discarded, then this must actually incline us to take the problem of crime and urban disorder more ser-iously and not less seriously. Thus, the question of whether anti-social behaviour is more prevalent now than before – while an important question – *is not the most important question* in this context.

Hence, it is argued here that the primary concern of historical con-tinuity analyses *should not be* about attempting to measure and to define anti-social behaviour as more or less pervasive than it has been in the past – with all the statistical and heuristic ambiguities that that embodies – it *should instead be* about identifying the way in which anti-social behaviour and its impact is being experienced differentially in *contemporary* Britain. And crucially, as we will consider shortly, to what extent social stratification and disadvantage can be seen as linked to disorder, and to what extent *personal responsibility* for anti-social behaviour should feature (Dennis and Erdos, 2005). Research suggests that anti-social behaviour is not experienced by the majority of the population and that it is a minority who are experiencing the most compounding and pronounced effects of anti-social behaviour, which, because of their economically marginal position, they are unable to escape from (Millie et al., 2005). The substantive evidence shows that the worst effects of anti-social behaviour are currently experienced dis-proportionately by a fewer number of people who, for the most part, are resident in social housing estates.

5
Anti-Social Behaviour and Social Housing

The use of anti-social behaviour orders and moreover, strategies of anti-social behaviour management in social housing areas has become one of the most contentious aspects of anti-social behaviour policy and practice. The use of anti-social behaviour interventions in social housing is often characterised as an oppressive and discriminatory process of social exclusion and control. As Atkinson (2006: 101) contends: 'Social housing has ... become both a site ... and process ... through which an urban poor are first concentrated, then managed and subsequently disciplined in line with the normative expectations of wider society given voice by a hostile, hysterical and sensationalist media'. Critical commentators contend that the deployment of anti-social behaviour strategies in social housing areas should be understood within the context of punitive networks of regulation, power and bureaucratisation. In this way, anti-social behaviour management is identified as a mechanism to more efficiently regulate the poor. Law, policy and other forms of bureaucratic control are thus utilised to control and to manage newly-defined anti-social behaviour(s) within these 'problem populations'. These critical analyses seek to represent the use of anti-social behaviour strategies in social housing as borne out of popular discriminatory proclivities coupled with an explicit/implicit desire to 'sidestep' fundamental problems of social structure emanating from economic deprivation and exclusion. While there are aspects of this perspective that are undoubtedly relevant *and* important to the debate on anti-social behaviour, it seems that critical analyses of this nature tend to neglect the importance – or even, simply, the relevance – of individual and community security as a 'thick public good' (Loader and Walker, 2007). Thus, in order that we might begin to consider more fully the arguments that underpin the academic debate on the use of anti-social behaviour strategies and ASBOs in social housing, let us now examine the nature and incidence of anti-social behaviour in social housing and the ways in which ASBOs are consequently being differentially applied in social housing areas.

Management and control in social housing

While Hester (2000: 172) predicted that ASBOs would be used primarily in 'poor communities' and 'by definition they will thus be disproportionately deployed', Brown has also argued that 'although crime is ubiquitous, anti-social behaviour is deemed to occur principally in social housing areas ... [which is] part of the broader social control of marginalised populations who can be "managed" in social housing' (2004: 204). However, as Flint (2006) observes, the 'policing' of the behaviour of tenants as an element of housing governance is not something new but in fact has a long historical precedent. Citing a paper written by the Chief Medical Officer in Glasgow in 1933 with the subject title *Can the Undesirable Tenant be Trained in Citizenship?*, Flint notes the obvious parallels with contemporary debates about anti-social behaviour, housing governance, and the role of the state and other agencies in affecting behavioural change in tenants. As such, Flint contends that the current policy on anti-social behaviour should be understood as an 'evolution' rather than a 'revolution' in approaches to anti-social tenants.

Indeed, Haworth and Manzi (1999) posit that moral assessments of tenants' behaviour has a strong historical continuity and that ethical processes have traditionally been used to determine whether tenants were 'deserving' or 'undeserving' of access to housing.

However, they also argue that such 'moral' assessments possessed a dual function: once tenants had been identified as 'deserving' of housing, their behaviour could then be subject to the normative constraints of tenancy agreements on acceptable conduct which could be enforced through a variety of applied sanctions. In a similar vein, Lister (2006: 125) suggests that the introduction of good neighbour agreements represents a departure from traditional relationships of trust and reasonable expectation between neighbours to a system of social control premised upon the creation of formal legal agreements between tenant and landlord.

Much academic scholarship is critical of the role of housing governance in 'managing' tenants' behaviour, in particular, through the use of tenancy agreements (see, for example, Saugeres, 2000). As Carr and Cowan (2006: 72) contend, tenancy conditions enable 'landlords to govern and responsibilise tenants, setting out norms of behaviour'. To what extent the right to housing should necessarily be *balanced with an obligation to respect ones neighbours and to behave reasonably* will be considered more fully in the course of this chapter. However, it is important to note that critical perspectives on housing governance often identify the normative regulation and disciplinary techniques invoked to constrain the behaviour of tenants as representative of a broader, overarching academic paradigm on the social control of marginalised populations and the authoritarian power of the state and its agencies. As Loader and Walker (2007: 82) contend, there is a deeply held, entrenched academic perspective on the role of the state in disciplinary practices which 'questions the sociological wisdom and normative value of [placing] such a deeply biased entity as the state at the heart of a project to produce more equitable distributions of policing and security resources'.

Additionally, critical perspectives also identify housing governance's perceived detachment from engagement with the socio-economic and socio-structural factors pervasive in anti-social behaviour cases. According to Card (2006: 50), there has been a linear historical preoccupation with ensuring the conformity of tenant/household behaviour which has been achieved, for the most part, in isolation from addressing (or even acknowledging) the 'wider social and structural explanations of their situation'. In fact, Card goes as far as to suggest that '...all council tenants have become perceived as irresponsible, work-shy and "undeserving"' (p. 54). Indeed, Card's somewhat zealous summation is that the popular perception of social housing (although she is not specific, one might assume that this would inculcate political, public and media spheres of opinion – although presumably not academic) is that 'Tenants of this tenure do not have the intellectual or financial resources to act responsibly and provide protection for themselves or their families against the risks encountered in life' (p. 48).

Similarly, Carr and Cowan (2006: 64) describe the entrenched and disconnected nature of the social position of council housing tenants as detached from the rest of society. They suggest that it is 'this division from the norm which justifies a certain racism and a form of killing, in the sense in which Foucault used those terms...And "the death of the other, the death of the bad race, of the inferior race (or the degenerate, or the abnormal) is something that will make life in general healthier: healthier and purer" (Foucault, 2003: 255)'. Reflecting upon anti-social behaviour policy as a means to segregate and to exclude, they suggest that it is unlikely that there has been a period since the creation of social housing that there has been a greater endeavour to demarcate the 'social' from the 'anti-social'.

These are, without doubt, essentially bleak analyses of the social housing tenant's position in society. What one must endeavour to determine is whether, stark as these characterisations may be, they are accurate in their representation of existing perceptions about and attitudes to social housing and also, if they are accurate in determining the status of tenants and households as disenfranchised, controlled, and monitored by an overarching punitive agenda of social control and social conformity dominated by elitist normative perceptions. That is to say, is there genuinely, as Card (2006) proposes, a pervasive view in society that 'all' council tenants are 'undeserving'? And, more importantly, are anti-social behaviour 'technologies' being expressly/impliedly used as a means to control and to punitively discipline social housing residents? Are anti-social behaviour strategies being *imposed* upon social housing areas or can such strategies also be understood in the wider context of genuine demands for action by local social housing residents?

A historical tradition of moral censure?

The majority of ASBO interventions continue to be issued in social housing areas (Home Office, 2006a; 2008; NAO, 2006; Scottish Executive, 2005b). This has been cited as evidence of a discriminatory agenda penalising non-conformity targeted at social housing residents. While Burney (2002) and Cowan et al. (2001) have observed the social housing sectors' increased use of and reliance upon procedures synonymous with crime control; Brown (2004) has commented that 'anti-social behaviour is found largely in social housing areas because the physical presence of "investigatory" people and technology ensure that it will be found' (2004: 210).

Yet, recent nationwide research on the incidence of anti-social behaviour has found that: 'anti-social behaviour has a significant impact on the lives of a minority of people in Britain, particularly in areas of social deprivation and inner cities. However, it has little or no effect on the lives of the majority of the population' (Millie et al., 2005: 1). Furthermore, 61 per cent of the respondents in the British Crime Survey (BCS) 2003/04 reported no negative effects from any of 16 types of anti-social behaviour (Home Office, 2004a). Both pieces of research were conducted nationally, so in these instances, anti-social behaviour was not found primarily in deprived urban areas simply because these were the only areas being studied. Both studies found anti-social behaviour to be present in other areas, however, Millie et al. found that anti-social acts affected the quality of life of residents to a lesser degree. Similarly, mirroring the findings of the Home Office study of *Crime in England & Wales 2005–06*, research on behalf of the National Audit Office (2006) also demonstrates that, although all research participants agreed that anti-social behaviour was a problem to some degree where they lived, participants from less affluent areas perceived anti-social behaviour to be a greater problem than those from more affluent areas (NAO, 2006: 9).

The financial costs of anti-social behaviour upon affected areas can also be considerable. Based on a 'one day count'[1] study of anti-social behaviour incidents in 2003, for example, the Home Office estimates that the 66,107 reports of anti-social behaviour recorded equate to £3.4 billion a year in associated financial costs (Home Office, 2006a: 28). This estimate does not, however, include the costs assumed by individuals. If costs attributed to others were included, the base line estimate for the cost of anti-social behaviour would raise substantially (NAO, 2006: 8). For example, the estimated annual cost to the victims of criminal damage is £1.2 billion (ibid.). Moreover, the Social Exclusion Unit reported in 2000 that the worst cases of anti-social behaviour can precipitate the demolition of recently built property and can result in the zero value of assets. It was estimated that the cost of demolition was approximately £5,000 per dwelling. However, this figure did not include the cost of re-landscaping the site and compensating previous tenants or owners (SEU, 2000a). Similarly, the associated cost(s) of vandalism to buildings is considerable: a study of the costs of vandalism in schools in Scotland, for example, estimated that the cost for insuring against vandalism and damage was higher than the amount spent on books each year (Accounts Commission, 1997).

In a study of social landlord's responses to anti-social behaviour, Nixon et al. (1999) reported that incidents of anti-social behaviour also have very high costs in terms of housing management time. The study estimated that 20 per cent of social landlords' housing management time was spent on dealing with complaints about neighbours' behaviour (Nixon et al., 1999). Landlords stated that tackling anti-social behaviour was a resource-intensive process which considerably impacted on housing management budgets. Research for the Scottish Office (1999) described these type of housing management costs as 'direct costs', that included the time spent dealing with neighbour complaints by housing officers, area managers, senior officers, and caretakers; the costs of implementing initiatives and associated on-going costs; legal costs for advice and court action; the associated costs of repairs for vandalism and graffiti; and the time given by homeless and allocation staff in dealing with requests for transfer.

Moreover, there are less likely to be amenities, services and shops in areas of high anti-social behaviour, primarily because of the associated costs of maintenance and repair (Brand and Price, 2000; Home Office, 2006a: 28). Research conducted by the Housing Corporation (1998) found that high levels of crime and anti-social behaviour in areas make housing difficult to let, reducing community participation, which subsequently leads to the rapid deterioration of these neighbourhoods. Communities with high levels of vandalism or graffiti can also discourage individuals from making use of community and neighbourhood areas as gathering places, which can have an affect on local business(es) as a result of reduced 'passing trade' (SEU, 2000a; Home Office, 2006a: 28). Additionally, Power and Mumford (1999) have found that low demand for housing in these communities subsequently generates falling school rolls, loss of confidence in the area, a vacuum in social control, increased anti-social behaviour and intense fear of crime.

The emotional costs to the victims of anti-social behaviour have also been reported for some considerable time. In 1999, for example, the National Housing Federation (NHF) found that tenants of social housing frequently suffered from high levels of stress as a result of crime and anti-social behaviour in the area in which they live (1999). Moreover, Upson (2006) found that 96 per cent of those suffering from noisy neighbours reported a resulting emotional consequence, which included annoyance, frustration, anger and worry. Of these respondents, 32 per cent detailed more serious emotional impact and disclosed having experienced one more of the following: shock, fear, stress, depression, anxiety, panic attacks and crying. Furthermore, anti-social behaviour adversely impacts on people's quality of life (NAO, 2006: 8) and victims may also suffer continued and prospective emotional distress caused by their experiences, such as depression and anxiety (Hunter et al., 2004).

This body of evidence, one could certainly argue, legitimately precipitates the use of interventions to address and to control acts of anti-social behaviour that are clearly so very detrimental to community life and to individuals' freedom from anxiety and sense of security. However, while research has evidenced the cost of anti-social behaviour (in terms of financial costs; community costs; and, costs to the individual) strategies of anti-social behaviour management continue to be critiqued on the basis that they are inherently 'moral' – in the sense that they are concerned with the erroneous identification and criminalisation of disadvantaged groups. Indeed, Howarth and Manzi (1999) view the evolution of social policy on housing and resident behaviour in recent years as part of a wider trend that entrenches 'an explicitly moral dimension into analyses of problems, with the result that increasingly punitive strategies are adopted.' Similarly, Waiton (2008: 340) raises concerns about the moral dynamic underpinning anti-social behaviour management as the new 'politics of behaviour'.

Interestingly, Waiton, citing Caldwell (1999: 5), suggests that there are parallels to be drawn with developments in the monitoring and regulation of behaviour in other fields. Using, Caldwell's analysis of the criminalisation of sexual harassment, Waiton considers whether the increased emphasis upon conduct and behaviour is representative of what Caldwell terms a 'dangerous shift in our understanding of civility [and] a tendency to inflate conflicts once resolved informally into wounding gladiatorial combat' (Caldwell, 1999: 5). Caldwell's analysis of sexual harassment and its 'criminalisation' strongly echoes a similar thesis on behavioural conduct advocated by Kors and Silvergate in their influential and strongly polemical book *The Shadow University* (1998). The authors contend that the infliction of a prescribed moral agenda through the use of behavioural codes in universities in the United States has resulted in the fundamental repression and subordination of individual liberty – in particular, the right to free speech. In this context, behavioural codes replace free speech rights in favour of a *predetermined notion of historical moral responsibility.* (see Kors and Silvergate, 1998).

The critique of this perceived infliction of a moral agenda through behavioural censure which Kors and Silvergate propose also extends to the area of sexual harassment. Using the behavioural code of Bowdoin College of Arts and Science in Maine as an example, the authors express incredulity at the range of prohibited sexual behaviours within the document, and its resulting flexibility. Yet, while the list of behaviours appear to Kors and Silvergate as ambiguous, and, one might ascertain, essentially not that 'serious', the creation of such codes was borne out of a legitimate need to protect vulnerable groups (in this case, women). The notion that a university or other institution should seek to prohibit men from leering, heckling and making inappropriate sexual remarks to women – which can be deeply distressing – is not in fact representative of a punitive moral agenda, but of a substantive need to protect women in the work place from harassment of a sexual nature which has been identified as a pervasive problem over many decades (Bimrose, 2004; Collins and Blodgeth, 1981; Farley, 1980; Mackinnon, 1979; Rowe, 1981).

The contention of Kors and Silvergate, and others who identify a 'moral agenda' on behaviour or indeed a 'dangerous shift in our understanding of civility' (Caldwell, 1999: 5), is that codes and other means of regulation should not be the vehicle used to enforce acceptable standards of behaviour. Rather, such critiques contend that individual and community accountability should be the predominant modes of ensuring civic responsibility. In this way, Waiton observes that:

> [A] framework is being established that encourages us all to resolve the irritations of everyday life, of noisy neighbours, rude commuters, rowdy kids and 'aggressive' customers, by contacting the growing array of authorities to deal with these problems for us. This discourages any possibility of social norms being established by the public itself and it also adds to the sense of individual impotence. Until recently, anti-social behaviour was understood as a problem to be resolved by themselves. When children swore and dropped litter or neighbours were noisy, people were expected to take a socially responsible approach and act themselves. ... Unfortunately, when we fail to take responsibility for these problems that, in our hearts, we know we should be doing something about, when we retreat into our bubbles, we diminish our sense of ourselves. (2008: 354–5)

There are two potential difficulties with this perspective. First, it could be argued that such a view necessarily understates the nature and affective dimensions of anti-social behaviour and how it is experienced and in essence, suggests that victims must *assume the burden* of resolving potentially dangerous or highly stressful situations rather than seeking help from the authorities. And second, proponents of anti-social behaviour policy might contend that this perspective overlooks the historical *ineffectiveness* of attempts made to tackle anti-social behaviour and neighbourhood disputes. In this respect, let us now consider the argument that there was in fact a *genuine need* for the imposition of a policy that sought to tackle anti-social behaviour in a more systematic (and potentially more punitive) way.

From the mid-1990s, the attention of both politicians and the media had been captured by particular communities in Britain that appeared to be wrought with problems associated with urban deprivation, social exclusion (Hills et al., 2002) and poverty (Gordon and Pantazis, 1997).

Residents and communities cited not only problems related to serious crime, but also the pernicious cumulative effects of anti-social behaviour and petty offending. However, as a result of the apparent impotence of civil law remedy in addressing neighbour(hood) nuisance, several councils became increasingly proactive in seeking greater powers to use against perpetrators of anti-social behaviour who were 'beyond the reach of both criminal and housing sanctions' (Burney, 2005: 20). For example, nuisance law had proved to be entirely ineffectual in *Hussain v Lancaster City Council [2000] QBD 1*, whereby a family had been severely and persistently racially harassed over a protracted period of time. The victim's claim against the local authority that they had failed to take action against the (tenant) perpetrators of said anti-social behaviour was struck out as disclosing no reasonable cause of action because the alleged anti-social acts were not committed from the perpetrators' land. It was held by Hirst L.J. at paragraph 23, that 'the acts complained of unquestionably interfered persistently and intolerably with the plaintiffs' enjoyment of the plaintiffs' land, but they did not involve the tenants' use of the tenants' land and therefore fell outside the scope of tort'.

Indeed, case law had shown there to exist intrinsic and fundamental problems for those seeking remedies as tenants. Where the anti-social act complained of was perpetrated by a fellow tenant of a common landlord, it had been established in *Hussain* that such a landlord would only be liable if he had *authorised* the acts of nuisance; this seemed to require that a let 'necessarily involved a nuisance' as per *Malzy v Eicholz [1916] 2 KB 308*. Moreover, the case of *Smith v Scott [1973] Ch. 314 321* not only indicated that this would be difficult to prove even if the landlord was aware of the troublesome nature of the tenants when housing them, but it also determined that there was no duty of care to existing tenants in the selection of new tenants. Accordingly, Brown has argued that previously, people had escaped conviction for anti-social acts 'for two reasons ... witness intimidation and ... the possibility that the police do not treat anti-social behaviour as "real" crime' (2004: 208). Similarly, Hunter et al. (2004) have further identified the long-standing lack of support available to witnesses in civil cases. Hence, during this period, the response of local authorities and landlords to neighbour complaints was clearly ineffective and deficient.

Subsequently, it can be argued that the current approach to anti-social behaviour was borne out of a *legitimate need* to protect vulnerable people and communities from the debilitating effects of anti-social behaviour in the areas in which the live. Not only was there evidence of the impact of anti-social behaviour on communities but there was also evidence that the law, as it was then constructed, was unable to deal effectively with (particularly persistent) acts of neighbourhood nuisance and urban disorder. It is unclear, however, what those who are critical specifically of the 'moral dimension' in anti-social behaviour policy, are advocating as an alternative means to addressing neighbour disputes. Nixon and Parr (2006: 96) state that the current policy on anti-social behaviour 'is likely to result in an increased unwillingness on the part of individuals to intervene. Rather than stimulate individuals to become active citizens contributing to the governance of communities in which they live, the political discourse may in practice engender further exclusions and divisions within already fragile communities.' Hence, while the overarching concern of the literature seems to be that neighbours should 'come together to confront some local blight' (Burney, 2005: 13), how this should work is practice and in what contexts it would be appropriate or effective has not yet been defined. Indeed, Waiton (2008: 355) concludes succinctly but ambiguously: 'the preoccupation with anti-social behaviour has emerged because of the loss of connection we feel with society and those around us. This is something that is being reinforced by an asocial elite who lack a social sense and are equally disengaged from "public" life'. What, then, are we to elicit from such an analysis of anti-social behaviour? Does the blame for anti-social acts lie with the economically/socially privileged (or even majority social groups) who have isolated themselves from community involvement and civic engagement? Are socio-economic structures, then, the most important facet of the anti-social behaviour paradigm? And how do we situate the notions of personal responsibility, accountability and 'blame' within the processes of anti-social behaviour management?

Responsibility and blame

One of the salient characteristics of academic literature on anti-social behaviour policy is the rejection of the construction of notions of 'blame' and personal accountability. That is to say, that much of the academic critique of anti-social behaviour management suggests that attempts to link the behaviour of anti-social behaviour perpetrators with moral choice is erroneous. This perspective is premised upon the belief that to equate blame is to ignore the impact of socio-structural factors upon marginal communities including, particularly, deprivation and economic decline. The idea that anti-social behaviour should be understood as (at least in part) a *moral* choice is characterised as reductive, limited contextually, and normatively suspect. As Squires (2008: 15, emphasis added) argues: 'the notion of anti-social behaviour has allowed left-realist ideas of causation to flip over, *holding individuals as responsible for anti-social behaviour,* and anti-social behaviour as the root cause of community decline. By definition, targeting anti-social individuals eclipses all other solutions'. In a similar vein, McIntosh (2008: 241, emphasis added) contends that the current approach results in '[b]ehaviour [being] *reduced to a moral choice,* crucially divorcing the importance of biographies, meanings and socio-structural contexts from how such behaviour must now be understood'. Moreover, critics contend that anti-social behaviour interventions, including the provision of support interventions for anti-social families and individuals, are exclusively propelled by a specific state sponsored *moral agenda* designed to 'regulate' non-conformist families (Gillies, 2005).

While it is necessary, no less essential, that we understand the influences and the difficulties affecting the perpetrators of anti-social behaviour, it could also be argued that the 'rejection of moral choice' argument that some opponents of anti-social behaviour policy have made, tends towards an over-simplified picture of anti-social behaviour perpetrators as straight-forwardly the victims of social processes such as urban deprivation and social exclusion. One of the very few academic voices advocating an alternative perspective in this area is Alan Deacon. In a theoretically informed piece on anti-social tenants, Deacon takes issue with the imbalance in academic literature which, he argues, consistently means that de facto concern about anti-social behaviour and its effects are outweighed by a supposition that initiatives aimed at tackling such behaviour will necessarily be discriminatory and oppressive (Deacon, 2004: 921). Instead, Deacon proposes that rather than viewing anti-social behaviour interventions to enforce civic obligations, and measures to reduce social exclusion and increase life chances, as antithetical and irreconcilable ideologically – they should be understood as 'two sides of the same coin' (Deacon, 2004: 912). Deacon presents a distinct challenge to the central tenet of the established literature which advocates that anti-social behaviour should not be identified as (simply) a moral choice.

Holt (2008: 208–9), for example, argues that the process of blaming the anti-social for their actions and their choice of behaviour 'operates within a landscape of morality where structural factors that may affect abilities to parent (such as housing, poverty and domestic violence) are ignored...', while Nixon and Parr (2006: 86) are critical of the 'consensus' that they suggest exists on behavioural accountability and which determines that individuals who refuse or fail to obey 'strict' rules and norms of behaviour must in turn be punished. Rejecting the inference upon individual 'choice', an 'advantage' of this technique of control, they contend, is that politicians are able to 'sidestep' the complicated issue of social exclusion. Nixon and Parr's use of punctuation marks around the word 'choice' infers that they believe that this is a subjective term. While behaviour should certainly be contextualised within the socio-economic environment in which it occurs, the structuralist nature of this paradigm presumably infers that there should be no *direct* correlation between action and choice. Naturally, the question must then arise, of when, if at all, are anti-social behaviour perpetrators deemed responsible for their actions? And does the 'rejection of moral choice' argument provide for an abdication of personal responsibility?

It appears that in the majority of the literature on the subject, no direct link is inferred between behavioural conduct and choice. For example, Lister (2006: 127) suggests that anti-social behaviour perpetrators might be unaware of existing legal frameworks and so consequently they may not be 'capable of adjusting their behaviour accordingly'. Given the ambiguous legislative definition of anti-social behaviour, and the very broad range of behaviour that might now reasonably be termed 'anti-social', this is certainly a possibility. Equally, however, one might consider that it is unlikely that very many anti-social behaviour perpetrators are not aware that vandalism, graffiti, drug dealing and property damage are illegal. Indeed, Deacon (2004: 918) discounts the argument that the conduct of anti-social behaviour perpetrators who live in social housing is a result of their disadvantaged position and instead he argues for a substantive link between conduct and choice:

> Those being blamed are not victims but people who have rejected the moral claims made upon them by and in the name of the community. They are subject to welfare conditions and sanctions not because of factors beyond their control but because of their own lack of civic responsibility. Those who argue that this represents the imposition of middle-class social norms fail to recognise the centuries old distinction between personal morality and civic morality. The duty to respect one's neighbour is quite properly a concern of public policy.

Moreover, a number of academic commentators are also highly critical of the language used to describe anti-social perpetrators. It is identified as discriminatory, connoting negative and value-laden characterisations of individuals who are already located within marginalised sections of society. As Nixon and Parr (2006: 91) contend, perpetrators 'are described in stigmatising, demonising and dehumanising language'. Without doubt, the use of language and discourse is extremely relevant to the debate on anti-social behaviour. While it is unobjectionably conceded that the categorisation of individuals and/or groups via terminology that connotes a negative depiction of their characteristics could, a priori, serve to influence the ways in which people interact with particular individuals/groups, it could also be argued that to suggest that *identifying* problem families consequently serves to 'create' or perhaps in some way to encourage the anti-social, is a tenuous link. It might be argued that such a view appears to go some way to excusing responsibility for anti-social acts, and that it appears to suggest that we should not identify and/or negatively categorise the anti-social for fear that a) we may create (more) anti-social behaviour and b) that the perpetrators will suffer negative consequences and their communities will judge them, misunderstand them, and relate to them differently. Of course, as Boas rightly observes: classification is not explanation (Boas, 1974 [1887]). Hence, in the same way that the term 'community' is unsatisfactory in terms of the diverse social contexts across which it is transferred (Stenson and Edwards, 2004), the categorisation of groups such as 'problem families' does not take account of mitigating circumstances such as addiction, mental health problems and learning difficulties – which are common features of ASBO cases (Brown, 2004). In this respect, there are undoubtedly legitimate discussions to be had about the need to balance enforcement measures with interventions of support.

In pursuing ASBO action, an applicant authority does not have to prove intention on the part of the defendant to cause alarm or distress. This effectively removes the requirement for criminal intent or *mens rea*, upon which criminal cases are dependent. Brown has argued that 'this explains why anti-social behaviour control is unconcerned about mental health problems, learning difficulties, addictions, domestic violence and other potential "mitigating factors" that are common features of anti-social behaviour cases' (2004: 206–207). Subsequently, there has been criticism, particularly from civil liberties groups and charities (BIBIC, 2006; Liberty, 2004; Mason, 2005; NAPO, 2004; SANE, 2005), of the use of ASBOs against the mentally ill, children with learning difficulties, peaceful protesters, those with problems related to alcohol/drug addiction, the homeless and prostitutes. For example, the Chief Executive of the mental health charity SANE, Marjorie Wallace, has stated that situations involving mentally ill people 'should not be allowed to degenerate to the point where the police become involved and an inappropriate course of action is taken in the form of an ASBO' (SANE, 2005). As a result, the inappropriate issuing of ASBOs has become an area of significant concern and debate which necessarily intersects with arguments about enforcement versus support within the anti-social behaviour paradigm.

This is undoubtedly an important area of discussion, particularly since prior to the introduction of the Respect Agenda (which placed a stronger focus upon support initiatives), little attention had been paid to preventing anti-social behaviour and to strategies of early intervention. However, as Mayfield and Mills (2008: 82) indicate 'The lack of a prevention focus on anti-social behaviour was the single most widespread concern among practitioners. The Respect Agenda...goes a long way to redressing the balance'. Similarly, Pawson and McKenzie (2006: 172) have found evidence of attempts by applicant authorities to acknowledge and to inculcate evidence of perpetrator vulnerability into their strategies of anti-social behaviour management. They note that: 'Contrary to the assertions of some critics, it does not appear to be the case that, in seeking to tackle anti-social behaviour, social landlords ignore the possibility of anti-social behaviour perpetrator vulnerability.' In this sense, there is now a greater implicit and explicit emphasis within anti-social behaviour policy (than there had originally been) upon applicant authorities to balance strategies of enforcement with interventions of support.

However, to what extent there exists an *individual* (moral) obligation – resulting from a fundamental form of civic responsibility – to behave 'reasonably' within one's community is highly contentious. Stephen (2008: 331) suggests that an improvement in community and neighbourhood relations cannot be actualised until socio-structural factors including, specifically, economic imbalance are properly addressed. From this will follow the 'enabling' of responsibility and the social embodiment of the derivative end product – 'liberty'. Stephen's argument is that security, in the sense of 'secure neighbourhoods', cannot be achieved *before* social inequity is addressed. In this sense, personal and civic responsibility are manifestations of egalitarian communities where social exclusion, economic marginalisation and profound disadvantage are no longer features. In a similar vein, Carr and Cowan (2006: 74) identify the 'extreme vulnerability' of anti-social behaviour perpetrators which they directly attribute to their 'marginal position' within society.

Thus, for many academic commentators, the political drive to tackle anti-social behaviour is misplaced, representing an opportunistic, populist agenda of criminalisation and net-widening which is apathetic about stimulating genuine, substantive action to change social conditions. Within this paradigm, political action for social change is often identified as of paramount importance. Such a view can be located as pervasive in the continuum of critical criminological scholarship which seeks to develop a historically grounded political economy approach that focuses on structured inequalities, in class, race and sex. Central to such critical thinking is the development of ideas and questions about criminology and social policy which focus on notions of social structure, economic inequality and social justice.

Repeatedly, we see the emphasis placed upon the need for social change (in terms of social structure) before 'respect' for one's neighbours and community can legitimately be required of perpetrators. Therefore, one must ask how far do we inculcate the notion of responsibility and personal accountability into our analyses of anti-social behaviour? Are economic and social disadvantage mitigating factors in accounting for anti-social conduct? The notion that anti-social behaviour is not a 'moral choice' (or any form of 'choice' at all) is frequently reproduced within the academic literature. Anti-social behaviour, we are reminded, mostly occurs as a result of inequity arising out of an individual's social circumstances.

Although much of their work has been primarily concerned with racial inequality, Blau and Blau (1982) contend that inequality in any form is a de facto potential source of conflict and violence. As such

> ... socio-economic inequalities undermine the social integration of a community by creating multiple parallel social differences which widen the separations between ethnic groups and between social classes, and it creates a situation characterised by much social disorganisation and prevalent latent animosities. (p. 119)

Indeed, Blau and Blau argue that inequality can be said to exemplify society's failure to make available adequate means for individuals to actualise culturally defined goals, illustrated by the 'prevalent disorganisation, sense of injustice, discontent, and distrust generated by the apparent contradiction between proclaimed values and norms, on the one hand, and social experiences, on the other' (ibid.). It follows that a deterministic view of anti-social behaviour which focuses upon social circumstances, necessarily rejects claims which seek to ascertain the primacy of personal responsibility for conduct.

However, returning to Deacon's theory (2004) on welfare provision and anti-social tenants, the obligation – or moral duty – to be a reasonable neighbour, should not be viewed as substantively contingent upon social renewal and efforts to address social exclusion, dependency and economic disadvantage. Indeed, Deacon posits that there is a legitimate argument to be made within social policy, including both housing and welfare, that programmes of neighbourhood renewal should not be seen as de facto precursors to tackling anti-social behaviour. In fact, Deacon (2004: 924) suggests that the concept that a right to housing should necessarily be 'balanced by an obligation not to abuse that housing is one that accords with basic sentiments of fairness and reciprocity'. Hence Deacon's central tenet is that there must be an implicit and overarching duty of fairness and reciprocity embodied within civic obligations. The duty to behave reasonably towards one's neighbours is therefore not dependent upon one's social or economic position but exists independently of social stratification. Citing White (1999), Deacon (2004: 924) maintains that:

> In the case of anti-social behaviour…the most compelling justification is to be found in the mutualist argument that public policy has to reaffirm and enforce the obligation to show respect and regard for the needs of others…economic redistribution will not automatically rebuild those 'bonds'. They have to accept that public policy has a further role in 'cultivating the disposition' of citizens to fulfil their obligations to each other and, 'in some cases, directly enforcing them'. (White, 1999: 167)

However, critics have argued that one of the fundamental difficulties with this concept of mutualism, and with the notion of shared values and respect for one's neighbours, is that modern society (and in particular, 'communities') no longer possess shared values. In essence, an individual's and a community's 'tolerance' of certain behaviour(s) is entirely relative. Jock Young (2001) proposes that criminal values may feature as an *element* of a given community as opposed to necessarily a negative power within it. Indeed, it has been suggested that particular communities may in fact be more 'tolerant' of anti-social behaviour occurring in their locales than residents living in other areas. The use of the word 'tolerance' here is an important choice of language. What is meant by this identified 'tolerance' to anti-social behaviour? For 'tolerance' can mean the capacity for, or the practice of, recognising and/or respecting the beliefs or practices of others but it may of course – equally – mean the capacity to endure hardship or pain.

Contexts of tolerance and intolerance

As we have already seen, empirical evidence has shown that anti-social behaviour has little or no effect on the quality of life of the *majority* of the population (Millie et al., 2005). Indeed, Millie et al. (2005: 1) found that anti-social behaviour in fact significantly impacts on the lives of a minority of people in Britain, and is concentrated in areas of inner city deprivation. Thus although anti-social behaviour should be seen as a problem affecting all housing tenures, it is those individuals who live in areas of concentrated multiple deprivation that are most likely to be the victims of anti-social behaviour and it is the residents of the most deprived inner city estates that 'tolerate' drug dealing and abuse, graffiti, vandalism and intimidation. The doctrine of cultural relativity carries most weight when argued by the people with the least power and opportunity to present their views – but we do not find that here. Research has shown that residents from the most marginalised and socially excluded estates experience more crime and anti-social behaviour and are more fearful of it occurring (Home Office, 1998, 2000a, 2003/04a; Millie et al., 2005; National Housing Federation, 1999; Power and Mumford, 1999).

As a result, it might be argued that the zealously propagated analysis of the social control element (in anti-social behaviour policy and legislation) has the potential to inaccurately reflect the substantive experiences of individuals and communities and moreover, the overarching requirement for 'security' as a 'thick public good' (Loader and Walker, 2007), illustrated in a body of research that demonstrates the 'stress' (NHF, 1999), 'intense fear' (Power and Mumford, 1999) and vulnerability (Home Office, 1992, 1998, 2000a, 2003/04b) experienced by those in areas of concentrated multiple deprivation. As Deacon (2004: 921) observes:

> The case for measures against those who perpetrate anti-social behaviour is further enforced by recent research on the nature of social capital. This has shown that what is important to people living in deprived areas is good neighbourly relations – the sense of security and well-being that comes from being able to call upon one's neighbour's in times of need.... It is this sense of security that is undermined by the anti-social behaviour of a relatively small number of households.

Arguments that suggest that forms of anti-social behaviour may feature as 'neutral' aspects of a community; and that particular communities/individuals are more 'tolerant' of anti-social behaviour than others reflect a popular overarching academic perspective on anti-social behaviour policy that, it might be argued, subverts the importance of the real consequences of anti-social behaviour on its victims. We are in a climate, Lianos posits, that defines 'research [on control] as a discourse that refers *as a matter of routine* to the erosion of liberties and the capture of society by dark and totalitarian forces' (p. 414, emphasis added). In this respect, it is crucial that analyses go beyond internalised understandings of anti-social behaviour, the forms it takes, and (perceived) levels of tolerance to it. In the context of this present discussion of 'tolerance' and 'intolerance' to anti-social behaviour, let us now endeavour to undertake a more expansive analysis of the behaviour that has been subject to censure through the use of anti-social behaviour initiatives, and in particular, through the use of ASBOs. More specifically, what does anti-social behaviour mean and to who? And how, if at all, is anti-social behaviour differentially interpreted in practice and, within the academic literature?

What is immediately apparent within academic writing is that there is a pervasive concern that individuals and communities are 'less tolerant' of nuisance behaviour than in previous decades. For example, Tonry (2004: 57) has argued that the elevated profile given to anti-social behaviour within political and media discourse has only succeeded in 'making more people aware of it and more dissatisfied with their lives and their government'. Pawson and McKenzie (2006: 173) claim that 'There is little evidence that anti-social behaviour is the growing menace sometimes portrayed by politicians and the media' and describe an 'associated inclination to "talk up" the issue as one demanding firm official countermeasures' (p. 155). From these perspectives, we can identify two salient and contentious issues related to academic understandings of anti-social behaviour and both relate to the concept of tolerance.

First, academic literature largely identifies anti-social behaviour as a problem with limited social impact which has, essentially, been exaggerated by political, media and popular discourses – and which has subsequently evolved into a more general and pervasive concern about incivilities and nuisance behaviour. Second, and perhaps even more importantly, some of the academic literature suggests that individuals and communities should be more 'tolerant' of nuisance behaviour because its substantive effects are minimal. Of particular interest to our current discussion is the discovery that some of the relevant academic literature on anti-social behaviour management views nuisance behaviour and incivility as, if not victimless, then certainly limited in its substantive impact upon others. For example, Burney (2005: 168) contends that 'There are no "victims" of litter, graffiti or passive gatherings of youths – just people who don't like these things or say they make them feel fearful'. In a similar way, Goldsmith (2008: 227) suggests that initiatives designed to understand the circumstances surrounding young people's anti-social behaviour are seen as reinforcing attitudes 'that they could *get away with it*. "It" was more often than not incidents of stone throwing, petty vandalism, noisy behaviour ...'. These two perspectives provide specific examples of behaviour that, it appears, are being identified as victimless or, by inference, not serious. Let us now deconstruct these examples and unpack the implicit assumptions underlying their use as examples of minor incivility.

Burney's assertion that there can be no 'victims' of littering or graffiti or 'passive' gatherings of young people, at first seems like a common sense assumption. However, as we know, anti-social behaviour is an elastic term with many variants. Similarly, littering, graffiti and 'passive' gatherings of youths have many incarnations. 'Littering' can range from fly tipping to dog fouling to dropping cigarette butts. It is harmful to wildlife, public health, and to the natural environment. 'Graffiti' is also a broad term and can include damage to property, including peoples' homes and businesses, vehicles, and public spaces. With regard to 'passive' gatherings of youths – we must consider what is meant by 'passive'? There is a pervasive concern within the academic literature that the identification of groups of young people (hanging around) as 'anti-social' is part of a broader agenda on the purification of space inculcating the 'demonisation' of youth and moreover, efforts to cleanse or 'sanitise' public space of difference. For example, Stephen (2008: 327) describes the targeting of young people in public spaces as part of the 'increasing criminalisation of public leisure'. But to what extent are efforts to prevent groups of young people gathering in public spaces an intrusion into their personal freedom to enjoy public space freely and could this type of intervention be considered duly warranted?

While groups of young people gathered in public spaces (and who are passive in the sense that they are not deliberately trying to intimidate others) is behaviour that should not require intervention, it has also been suggested that groups of young people gathered outside the local shop or supermarket, who make insulting comments, or who heckle and jeer passers by – but do not actively engage in any substantive criminal behaviour – should not be subject to anti-social behaviour interventions which may ultimately have criminal law consequences (such as the ASBO). This view appears to be contingent upon a particular perspective on 'tolerance' – and by implication, what types of behaviour individuals should be *expected* to tolerate. Goldsmith's example of 'stone throwing...' is also contentious. It could be argued that the inference is that the range of behaviours cited are not particularly serious and in principle may not be worthy of such community 'outrage'. But if we deconstruct 'stone throwing' as just one example, what are its consequences upon others? Personal injury? Destruction of or damage to property? Financial loss? Fear? Stress? Proponents of anti-social behaviour policy might argue that a civil order (with criminal sanctions upon breach) prohibiting stone throwing is an appropriate means to deal with behaviour of this nature.

We must also consider the broader question of how academic critiques respond to and represent evidence on anti-social behaviour and the use of ASBOs. That is to say, how does academic literature present the views of perpetrators of anti-social behaviour, and how does it present the experience of victims of anti-social behaviour? For example, Burney (2008: 146), citing the research findings of Wain (2007), describes how the participants of his study found the conditions of their ASBOs 'most irksome' and how 'several interviewees told of bad effects on their families emotionally and socially ... others spoke of the way their ASBO had tarnished their whole family following publicity'. It could perhaps be argued that the tone appears, to a degree at least, sympathetic towards the participants, and the negative and/or restrictive effects that the ASBO prohibitions have had upon their lives. Indeed, one might observe a sympathetic tone in other academic accounts of anti-social behaviour perpetrators' experiences. For example, Squires (2008: 24) suggests that young peoples' characterisation of ASBOs as 'badges of honour' might be attributable to the perpetrator's marginalised social status: 'how else could they respond: a defiant, self-destructive, masculine bravado in the face of hopelessness, the only resource left?'

Similarly, in their research, Nixon and Parr report on anti-social behaviour perpetrators' beliefs that the claims made against them were either wholly, or in part, illegitimate: 'They reported that the behaviour being complained about either did not happen, was exaggerated or was simply behaviour that is common to many families' (2006: 94). And in their interviews with victims of neighbourhood nuisance they observed that

> ... neighbour disputes often become a site where identities were established through a binary classification of 'them' and 'us', where the 'other' are the dysfunctional, 'irresponsible' and 'selfish' and

'we' are the 'normal', 'decent' and 'hard-working'. At no point was there an acknowledgement of the socially constructed nature of the problem ... (ibid. p. 80)

While Nixon and Parr (2006: 85) contend that 'the political discourse is dominated by otherness and exclusion, a lack of specificity and a failure to reflect on the complex way in which the role of the victim and perpetrator can be interchangeable', proponents of anti-social behaviour policy might in turn argue that academic discourse is dominated by a concern with social stratification which has the potential to represent the experience of 'victims' of anti-social behaviour in a reductive and marginalising context – which fails to accurately take account of the affective dimensions of neighbourhood nuisance. In this way, academic literature might be criticised for becoming too embedded in its own theoretically informed notions of social stratification and failing to understand the substantive circumstances informing people's conceptualisations of anti-social behaviour perpetrators as deviant 'others'.

While some commentators have advocated a relativist approach to understanding anti-social behaviour that is premised upon an acknowledgement that culture is an essential factor in the study of anti-social behaviour, Burney (2005: 9) observes that '... poor people do suffer more from crime and disorder but they also have more things to worry about and are more likely to feel things are out of control'. Both arguments are premised upon the notion of cultural relativism but are also linked closely with Cohen's theory of 'moral panic'. Indeed, Burney (2005: viii) contends that anti-social behaviour should be understood as 'a pyramid of different kinds and degrees of incivility, where only the tip is the painfully sharp experience, compared with a wide base of mere distaste or annoyance'. Fear about anti-social behaviour is often represented as *exaggerated*, the inference being that concerns about crime and nuisance which are located within areas of economic deprivation ought to be interpreted within the wider context of the compounding difficulties experienced by individuals resident in these areas.

However, Dennis and Erdos make an important observation when they note that, given the existence of empirical evidence which has found that anti-social behaviour is more prevalent in areas of concentrated multiple deprivation, the traditional assumption of the 'moral panic' advocates – that anti-social behaviour is generally an exaggerated concern of the middle-class (often founded upon notions of 'cosmopolitan intolerance' and 'middle-class sensibilities') – cannot be validated. As such, they argue that

> …if or to the extent that fear of crime is indeed exaggerated, the figures…directly contradict the 'moral panic' theory that it is respectable society (meaning the well-off) that does the exaggerating. These data show that if or to the extent that the fear of crime is exaggerated, it is the poor who do the exaggerating, not the rich. (2005: 65)

Similarly, it might be argued that it is somewhat tenuous to suggest that because 'poor people' have 'more things to worry about' that anti-social behaviour is likely to impact upon them more seriously (Burney, 2005: 9). Not only does this appear to diminish the *legitimate* concerns and experiences of people affected by anti-social behaviour in social housing areas, it suggests that it is *only because* these individuals have other difficulties in their lives that they find acts of anti-social behaviour so problematic. The implication is that if the specific behaviour occurred elsewhere (in a middle-class area) that the consequences would not be felt so abruptly. Such a view appears to discount the possibility that the empirical evidence on the higher incidence of anti-social behaviour in social housing areas is substantive rather than culturally relative.

It is worth noting that Dennis and Erdos have also rejected related claims that higher levels of 'fear of crime' experienced (particularly) in areas of concentrated multiple deprivation are based upon illogical anxieties as opposed to being grounded in the genuine prospect of crime or disorder occurring. They summarise as follows:

> To dismiss the 'fear of disorder' or the 'fear of crime' as something irrational, and to say that one of the jobs of the police is to talk old people in particular out of their allegedly irrational fears, is to miss the point. Logically, the implied argument is absurd, namely, that if only one in a thousand old people is verbally abused or mugged, only one in a thousand old people should fear being verbally abused or mugged. Empirically, the level of harassment or crime against old people is kept down precisely by their fear of being the victim of harassment or crime. Young men are subject to violence more than old women at night, partly because more young men are in the pubs. Old women are less likely to be the subjects of violence because they confine themselves to the comparative safety of their home more often than they once did. (Dennis and Erdos, 2005: 176)

Indeed, it is the broad contention of this book that anti-social behaviour – and fear of anti-social behaviour – should not be minimised or discounted or seen as exaggerated. It is a genuine substantive problem, which is experienced disproportionately in areas of concentrated multiple deprivation. Further, although anti-social behaviour is a subjective term, it is problematic to understand anti-social behaviour as entirely culturally relative – in the sense that poorer communities might be more tolerant of it occurring. A cultural relativist position in this sense is more likely to entrench the marginalised status of victims of anti-social behaviour. Of course, critics have argued that the notion of a 'common good' or 'shared values' is impossible in a modern, pluralistic society. Moreover, it has been suggested that the notion of civic reciprocity is naïve in its understanding of power/structural relations which impact upon which values/standards become entrenched in a community (Collins and Cattermole, 2006). However, in the same way that law inculcates an objective standard of reasonableness (the 'reasonable person') by which conduct can be measured, it is also possible to measure the notion of civic responsibility in this way. The duty to behave *reasonably* towards one's neighbours is not so subjective as to defy classification. For many academics, however, the classification of anti-social behaviour as conduct causing (or likely to cause) harassment, alarm or distress is unduly wide and has in practice resulted in the criminalisation of nuisance and a punitive focus on incivilities. As such, they are critical of what is seen as the criminalisation of conduct that would not traditionally have attracted the attention of the criminal law. Let us now consider these criticisms in more detail by turning to an examination of ASBOs and the court system.

6
ASBOs in Practice

The study of ASBOs in Britain requires that attention be paid to the social factors underpinning their introduction and use, but equally, to the legal and court process(es) that intersect to shape practices and outcomes. In previous chapters, we have considered the socio-political and historical background to the creation of ASBOs. Drawing in part upon my own empirical research, what now follows is a discussion of the *structure and organisation of the law governing the use of ASBOs*, and the formal role of legal procedure(s) in determining processes and substantive outcomes. Here we will consider how ASBO powers are administered in practice, and to what extent legal provisions are effective in limiting and regulating their use. As Squires (2008: 17) portends, 'the most important legacy of anti-social behaviour could well be the way in which it has become the foundation upon which a whole new range of hybrid, and semi-criminal, enforcement powers has been brought into being'.

The study

The subject of my research study (hereafter referred to as the 'Study') was a socio-legal analysis of the administration of ASBOs, and the ways in which the dimensions of due process; and juridical power and discretion intersect to shape the management and outcomes[1] of ASBO use in Britain. Despite fundamental differences in the legal and institutional systems of the three jurisdictions studies (England and Wales, and Scotland), the statutory similarities within the Anti-Social Behaviour Act 2003 and the Anti-Social Behaviour Etc. (Scotland) Act 2004, and the shared policy on countering anti-social behaviour, mean that data derived from each jurisdiction is both appropriate and relevant to research on the use of ASBOs as a whole. However, *the findings from the research that are discussed below are not presented as directly or substantively comparable with its composite jurisdiction north/south of the border.*

Sample and data sources

The data production approach(es) applied in the study were both quantitative (positivist) and qualitative (phenomenological) in their composition – hence, a pluralist (mixed method) research design was used. Quantitative data was obtained through the use of an online survey questionnaire, and qualitative information was derived from both 'unobtrusive' (Lee, 2000) methods of data collection (document examination), and semi-structured interviews.

A pilot study was conducted involving face to face interviews with five anti-social behaviour unit managers in England and Wales, and three anti-social behaviour unit managers in Scotland in order to identify salient themes for the study proper. The main study used survey data collected from solicitors involved in *pursuing ASBO applications* in England and Wales, and in Scotland. As the contact details for local authority solicitors are not freely available, it was necessary to first approach individual local authority anti-social behaviour co-coordinators/community safety officers.[2] Contact details for anti-social behaviour co-coordinators/community safety officers were available on local authority web sites and also on the Home Office's 'Together' website. The relevant officers of each local authority were contacted directly, giving them details about the study. They were then asked if they would consider forwarding an email to their solicitor(s) involved in ASBO applications, detailing the nature of the research project and providing them with the link to the online survey questionnaire. Many local authorities in England and Wales are not involved as the lead agency in pursuing ASBO applications for their area. Instead, ASBO applications are exclusively applied for by the local police force.[3] In these instances, the local authority anti-social behaviour co-coordinators/community safety officers provided contact details for the relevant local police officer, who was then contacted with the same details about the study as previously detailed.

A number of the larger local authorities (for example, Manchester city council) use an external firm of solicitors for their ASBO cases, and do not have an affiliated local authority internal solicitor(s) for ASBO applications. In these instances, contacts were provided for external solicitors who were approached to see if they would consider taking part in the study. The survey responses also include those from individuals within the Crown Prosecution Service (CPS) who were involved in seeking orders on conviction, although this accounted for only a very small proportion of potential respondents (2 per cent). In order to differentiate between the numbers of responses provided by an internal solicitor (specific to one local authority) and the number of responses provided by an external solicitor/CPS (who can represent multiple authorities), the survey questionnaire asked respondents to identify whether they acted in the capacity of internal or external counsel. However, it was not possible to determine how many solicitors from each local authority had answered the survey questionnaire. It should also be acknowledged that, because local authority staff had access to the survey URL, they were also potentially able to respond to the survey themselves (although it was specifically communicated to them in the email that the survey sought responses from solicitors only).

The survey was completed by respondents (in all jurisdictions) between January and March 2007. In Scotland, of the 32 local authorities approached, 18 survey responses were completed and analysed. In England and Wales, of the 410 local authorities that were approached to participate in the study, 137 survey responses were completed and analysed. These responses include those from police force solicitors involved in ASBO applications, where the local authority is not involved in ASBO applications for the area.[4] It should be noted that Registered Social Landlords (RSLs) are also empowered under the relevant legislation to make ASBO applications.[5] However, research has shown that the number of ASBOs originating directly from RSLs is small. For example, in Scotland, 13 per cent of full ASBOs were found to originate from RSLs/housing associations/co-ops, (Scottish Executive, 2005a: 4.1). Due to the high number of RSLs in existence in Scotland (296), and in England and Wales (over 1,800 in England alone), it was felt that it would not be prudent or expeditious for the purposes of this research study to contact RSLs to try to obtain contact details for solicitors involved in ASBO applications that they may be pursuing.

Moreover, no defence solicitors were approached to participate in the survey. The decision not to survey defence lawyers was made for two reasons. First, in order to legitimately compare the survey responses of prosecution lawyers with defence lawyers, broadly similar sized samples would be required. It was felt that it would be difficult to obtain the contact details for many defence lawyers involved in ASBO cases as these details are not necessarily freely available and would involve a level of 'cold-calling', which may have been viewed as an invasion of privacy (Schillewaert et al., 1998; Swoboda et al., 1997). Second, another survey would have had to have been constructed for defence lawyer participants, and a new data set(s) created. As such, it was felt that the time constraints of the study meant that this would not be prudent. The study of defence lawyers in ASBO applications would, however, be a particularly useful contribution for future research in this area.

The study also used data obtained from 11 semi-structured interviews conducted with Judges (Sheriffs) deciding on ASBO applications in the Sheriff Courts in Scotland which yielded a significant – and valuable – amount of empirical study data. The interviews were conducted between April and June 2007. It had been the aim of the study to conduct parallel interviews with Judges in the courts in England and Wales. However, the study access request was denied by the Senior Presiding Judge on the grounds that it would be inappropriate for interviews to be conducted with district judges and magistrates who would be asked to comment on the decisions of the higher courts. Gummesson (1991: 21) argues that gaining access to participants is the researcher's single biggest difficulty – which proved to be true in the context of this study. Other researchers have found similar difficulties in attempting to access the courts – for example, Ashworth et al. were denied access to the judiciary for their study of sentencing in the Crown Court (1984), while Hood experienced similar difficulties in his study of racial disparity in sentencing (Hood, 1992). However, it is unfortunate that access was denied to the lower courts in England and Wales, particularly because there exists so little research on judicial decision-making in ASBO applications and so the proposed interviews would have been timely, and would, moreover, have contributed significantly to knowledge in this area.

Data limitations

The research findings have specific limitations which impact upon the discussion of the study data. Due to the limited amount of data obtained from the (18) survey responses in Scotland, it was not possible to arrive at robust findings and generalisations about legal procedure in ASBO cases in Scotland. Consequently, where legal procedure is discussed below using survey data, it is discussed only with reference to data obtained from the (137) survey responses in England and Wales. Similarly, where reference is made to data obtained from the study interviews with the judiciary, discussion will be informed by data obtained only from the Sheriffs in the Scottish Courts since access was denied to the judiciary in England and Wales. *Hence, the findings from the research that are discussed below are not presented as directly or substantively comparable with its composite jurisdiction north/south of the border.* Instead, it is hoped that the findings presented will further illuminate understandings of ASBO legal and court procedure within the jurisdictional areas studied, and moreover, that given the shared policy on countering anti-social behaviour in the three jurisdictions studied, the data provided below will highlight areas that may inform or may be of interest to future research in other jurisdictions.

Civil procedure

The ASBO is designed to be a *preventative* remedy and not a punitive sanction. Thus, the civil law status of the orders has implications for the type of court proceedings at which applications are heard. In considering the classification of the orders in the landmark House of Lords case *R (McCann) v Manchester Crown Court [2003] 1 AC 787*, their lordships confirmed that ASBOs are civil orders and made ASBO applications an exception from the normal standard of proof in civil proceedings (on the balance of probabilities), ruling that the heightened civil standard, equivalent to the criminal standard, was to apply.[6] The civil classification of the ASBO process means that civil rules of evidence apply, including the use of hearsay and professional witness evidence. However, although the application for an ASBO is a civil process, the consequences of the breach of an order are *criminal*. This, in turn, has implications for the burden of proof in ASBO cases: ASBO proceedings are subsequently regarded as quasi-criminal in nature (*R (McCann) v Manchester Crown Court* [2002] All ER 593). It was argued by the defendants in *McCann* that, when considering the appropriate legal status of an ASBO, the courts should have regard to the proceedings leading to the imposition of an order, *but also* that the court should acknowledge that criminal proceedings may be brought if the order is subsequently breached. Lord Steyn (at [23]) set aside this argument:

> These are separate and independent procedures. The making of the order will presumably sometimes serve its purpose and there will be no proceedings for breach. It is in principle necessary to consider the two stages separately.

Further, in *Clingham v Kensington and Chelsea RLBC [2003] HLR 17*, where the imposition of an ASBO was challenged as being contrary to the Human Rights Act 1998, Lord Steyn described the balancing of rights in ASBO cases thus:

> The view was taken that the proceedings for an anti-social behaviour order would be civil and would not attract the rigour of the inflexible and sometimes absurdly technical hearsay rule which applies in criminal cases. If this supposition was wrong, in the sense that Parliament did not objectively achieve its aim, it would inevitably follow that the procedure for obtaining anti-social behaviour orders is completely unworkable and useless. If that is what the law decrees, so be it. My starting point is, however, an initial scepticism of an outcome which could deprive communities of *their* fundamental rights…. (at para 18; original emphasis)

The argument that the (potentially) onerous conditions that could be attached to an order necessitated that the proceedings be regarded as criminal was thus rejected. In support of this decision, Lord Steyn also described the draconian nature of many civil law injunctions such as *Mareva* injunctions[7] and *Anton Piller* orders,[8] and indicated that ASBOs should be considered in the context of facilitating protecting the rights of the *community*.

Jurists and academics have repeatedly argued that the 'amalgamation' of elements of the civil and criminal law within the relevant anti-social behaviour legislation has effectively blurred the 'fundamental boundary' between the civil and criminal law (Burney, 2002: 483). By combining civil rules of evidence with a criminal standard of proof, critics have suggested that not only is the distinction between the civil and criminal law obfuscated, but that proper procedural safeguards are circumvented. That the ASBO does not pay enough attention to due process is one of the central criticisms of its use.

The criminal justice models of England and Wales, and Scotland[9] are, despite structural variations, both underpinned by the overarching principle of due process. That is to say, both systems of law embody procedural protections that are intended to apply to defendants without regard for the seriousness of the penalty faced by the defendant. However, Ashworth and Zedner (2008) contend that the traditional safeguards of due process associated with the Anglo-American criminal trial are being eroded, resulting in the disintegration of established concepts of procedural justice (see also Ashworth, 1998; 2004), through, in particular, an escalation in the use of summary trials and the increased use of hybrid civil/criminal remedies – specifically including the ASBO. As such, a significant body of socio-legal and criminological writing has developed in the last few years which has argued that the use of ASBOs in Britain cannot satisfactorily be reconciled with the democratic justice principles of the Anglo-American criminal justice model. As identified above, scholarly critiques centre upon the legal inventiveness of the ASBO as a hybrid civil/criminal order, and the derivative due process issues arising from its creation. The evidentiary requirements and the applicable standards of proof arising from the various types of ASBO action[10] are distinguished as specific issues of concern. Simester and von Hirsch (2006: 192), argue that ASBOs have 'the potential to visit significant deprivations without adequate notice and procedural protections', while highlighting the more pervasive, recurring concern that ASBOs can be used to target 'undesirable persons' (p. 181).

In his discussion of the human rights element in anti-social behaviour legislation, Ashworth (2004: 268) posits that the creation of ASBOs as a *preventative* remedy in civil law has been 'an attempt to take maximum advantage of legal forms', which essentially enables relatively oppressive conditions to be attached to orders by virtue of their civil law status. The 'balancing of rights' is seen as particularly problematic in this context, since Ashworth maintains that the 'public interest' has been used by the courts as a justification for discarding human rights protections. And it is certainly the case that the courts have sought to prioritise the 'rights' of communities in ASBO action. By way of illustration, in a consideration of the use of hearsay evidence in ASBO applications, and the need to balance the rights of the defendant and the rights of the victim, Lord Hutton argued thus:

> I consider that the striking of a fair balance between the demands of the general interest of the community (the community in this case being represented by weak and vulnerable people who claim that they are the victims of anti-social behaviour which violates their rights) and the requirements of the protection of the defendants' rights requires the scales to come down in favour of the protection of the community and of permitting the use of hearsay evidence in applications for anti-social behaviour orders. (*McCann* at [113])

Many critics of the ASBO model identify this notion of 'balancing rights' as particularly problematic. Indeed, as Zedner (2005), amongst other criminologists, has argued the experience of criminal justice is that 'balancing' is a politically dangerous metaphor unless careful consideration is given to what (and whose) values/interests are involved. Given that ASBOs (and anti-social behaviour interventions more broadly) are most often administered in areas of socio-economic disadvantage, it follows that analyses should necessarily endeavour to objectively consider which rights are at stake, in what instances they are surrendered, and for what (and whose) gain.

As Loader and Walker (2007: 54–6) have observed, liberal thought has, historically, been very apprehensive about the idea of balance, and how it should be applied, in the context of security. They summarise the source of this concern as follows:

> ...liberals tend to be deeply uneasy about the invocation of conse-quential reasons for the erosion of liberties. ... Our most fundamental freedoms, whether of expression, assembly or liberty of the person, are considered too precious simply to be weighed on a social scale against competing goods and diminished in proportion to the inten-sity of the claims made on behalf of these competing goods ... the moral instinct of the liberal is typically to give some special status (even if not an absolute priority) to the protection of basic liberties, which means that they cannot simply be traded away for other social gains. ... Balancing, in short, promises a false objectivity and suggests an artificial precision in our attempts to devise just institutional arrangements and answer liberal concerns ... the idea of balance can provide no guarantee against structural tendencies deeply inscribed in the social foundations of policing and security.

With regard to ASBOs specifically, the idea of balance is thus seen as problematic because it appears to assume that public interest can take pre-cedence over fundamental human rights (Ashworth, 2004). Collins and Cattermole (2006: 32) have identified that this form of juridical rationale is not peculiar to ASBOs. Citing *Brown v Stott* [2003] 1 AC 681, where a woman challenged her conviction for unlawful driving on the basis that it breached Article 6 (right to a fair trial) of the European Convention on Human Rights, they observe that the court (erroneously) found that public interest in effect 'trumped' human rights considerations.

Moreover, the notion of balance in ASBO action is also critiqued on the basis that it is invoked to protect the interests of majority social groups, in effect facilitating the suppression and domination of marginalised and socially excluded groups. Carr and Cowan (2006: 69) have critiqued the use of civil rules of evidence in ASBO cases, suggesting that

> ...the dismantling of traditional restrictions on the use of anonym-ous evidence becomes inevitable and unchallenged. This operates to

exclude the anti-social from the normal and oratorically universal protections of the law. Common sense therefore justifies the death of the social existence of the 'other' because of the need to enhance protections of the 'innocent' and 'the law-abiding'.

Carr and Cowan (2006: 68), however, go further still. Using Lord Steyn's opinion in *Clingham* that a balance must be struck so that communities are not deprived of *their* rights, and then his lordship's observation in *Manchester City Council v Lee [2004] HLR 11 161* that anti-social behaviour constitutes a 'social problem', they suggest that Lord Steyn is guilty of judicially 'ventriloquiz[ing] the "national" community'. Indeed, they assert that 'what Lord Steyn is doing...is quite different from the normal legislative function of judges' (p. 68), and they argue that such 'discursive strategies...have a kind of negative encoding in which important elements are hidden away in what is left unsaid' (p. 69).

In this way, critiques of the ASBO process are asserting that we ought to consider how law is being used as a mechanism to protect and to guarantee liberties, but that also, we should ask *whose interests* are being protected in the context of power relations and social stratification. The suspicion is that ASBOs are being used predominantly to control and to exclude individuals from already socially or economically marginalised positions in society, which is part of a broader 'culture of control' that places an increased focus on victims, which normalises crime, and which encourages greater community involvement in justice processes (Garland, 2001). As we have seen in the previous chapters, these concerns are pervasive throughout the academic debate on anti-social behaviour management. The management of anti-social behaviour is identified with a historical continuity of (social) housing regulation, and criminalisation of the poor/powerless, which is now being facilitated through the decline of the criminal law and the increased use of hybrid orders such as the ASBO (Ashworth and Zedner, 2008).

Yet despite the substantial criticisms made of the ASBO model by its opponents, it is argued here that the quasi-criminal nature of ASBO proceedings (in the respect that proceedings are civil, with civil rules of evidence, and a criminal standard of proof applies) *is not necessarily problematic in and of itself.*[11] However, the classification of ASBO proceedings must certainly be examined in terms of the *substantive operation* of the relevant legislation, and arguments about the civil law status of the ASBO should, at least in part, be investigated from a functionary perspective. That is to say, while the sociological contexts of anti-social behaviour management are relevant, equally as are the legal contexts – and in particular, how the law is operating *in practice*.

The 'hybrid' nature of civil and criminal procedure seems to have generated a degree of confusion and practical difficulty within the court process in ASBO applications. (Indeed, in her early research on ASBOs, Campbell (2002: 49) appears to have confused the 'civil' nature of the orders with a 'civil burden of proof'. She cites *McCann* as settling the issue on the burden of proof required – although she misunderstands that it is the heightened civil standard, and not the traditional civil standard – 'on the balance of probabilities' – that applies.)

In the aforementioned Study of solicitors' opinions on ASBO legal and court procedures, nearly a quarter (24 per cent) of solicitors in England and Wales cited difficulties relating to court procedure in ASBO applications. The specific problems encountered were described as magistrates' court procedure; a lack of guidance on the level of evidence required for interim ASBOs; confusion regarding the 'necessity test';[12] difficulties relating to the burden of proof in the county courts (due to the two different standards of proof that applications are required to meet – the standard of proof for the ASBO and the standard of proof for the action that the order is ancillary to); and, a frustration that the majority of ASBO cases could only be heard in the magistrates' court. When asked about possible improvements that could be made to current legal and court process(es) in ASBO cases, the majority of solicitors (62 per cent) cited the use of the County Court for (stand alone) ASBO applications as a major improvement that could be made to the existing court procedure for ASBOs. It was suggested that because ASBOs are in many ways akin to anti-social behaviour injunctions (ASBIs),[13] County Court District Judges and Officers would be more able to process such cases expeditiously (unless the order is made on the back of a conviction). Moreover, many respondents found the Magistrates' Court process cumbersome when compared with that of the County Court. It was suggested that the use of the County Courts would benefit applicants and defendants if proceedings for stand-alone ASBOs were made in the County Court.

However, nearly two-thirds of respondents (65 per cent) did not find it problematic that civil rules of evidence are used in ASBO cases when (the equivalent of) a criminal standard of proof applies. The predominant reason given by solicitors for why they found civil rules to be unproblematic (given the equivalent of a criminal standard of proof) was *the importance attached to the use of hearsay evidence* in ASBO applications. Solicitors who supported the use of civil rules of evidence in

ASBO applications, described the use of hearsay evidence as 'vital' and 'crucial' in the ASBO process, primarily with regard to protecting vulnerable witnesses who would not otherwise testify in court for fear of reprisals. For instance, 68 per cent of respondents in England and Wales reported obtaining interim ASBOs based only on hearsay evidence and 22 per cent of solicitors had also been able to obtain a full ASBO in this way. It is important to note, however, that there is no statutory requirement that evidence should be led at the interim stage and there is no explicit provision for any representations to be made by or on behalf of the respondent before an interim ASBO is granted, although the Court can consider any such representations as it sees fit. As such, in England and Wales, an interim order may be granted *ex parte*, without intimation to the defender, and without any defence(s) having been lodged or presented in court.

These findings potentially raise questions about the legitimacy of legal action following breach of an order obtained solely on the basis of hearsay evidence. Although it was observed in the English case of *R (Keating) v Knowsley Metropolitan Borough Council [2004] EWHC 1933 (Admin)*, that, where a court is concerned with interim proceedings, it must bear in mind that no findings of fact have been made, that any allegations have not been proved, and that the defendant has had no opportunity to challenge the allegations, solicitors in the Study argued that, where an interim order has been granted on the basis of hearsay evidence and the defendant is subsequently arrested for breaching the order, the breach *should not* carry criminal sanctions. One solicitor described their opposition to criminal sanctions for interim orders thus:

> I feel that it is highly unfair that an interim ASBO can be granted without the need for any evidence to be led and then that an interim ASBO can lead to a criminal conviction. The government is wanting its cake and eating it. They say the interim ASBO can be granted without the need for evidence because it is a civil order designed as a deterrent, but then people are being arrested for breaching an order, the validity of which has never been tested in court. I do not feel interim ASBOs should carry criminal sanctions, it is oppressive.

In this context, it is important to note that one of the main reasons for the creation of the ASBO was as a means to tackle neighbour disputes whilst providing protection to witnesses who were vulnerable to intimidation and acts of retribution. Moreover ASBOs were also seen as a device to tackle the problem of domestic violence in those cases where women were too frightened to give evidence in court against their attackers.

Hence the hybrid process enabled use of hearsay and professional witness evidence to be used in ASBO action in order to protect vulnerable witnesses. However, the use of hearsay has not entirely extinguished the difficulties associated with witness (fear of) intimidation. In the aforementioned Study, survey respondents described the difficulties that they had experienced in the evidence-gathering process, particularly with regard to obtaining testimony from witnesses vulnerable to intimidation or acts of retribution. The majority of respondents (71 per cent) had experienced problems in securing witnesses, of which 95 per cent identified this as directly attributable to witness 'fear of reprisals'.

Although the use of civil rules of evidence was designed to enable the use of hearsay and professional witness evidence to protect vulnerable witnesses, it appears that the use of civil procedure has not produced the intended protection for witnesses and that some witnesses continue to suffer intimidation and retribution (before, during, and after) the court process. Solicitors in England and Wales identified a range of problematic areas in the use of witnesses. Indeed, they cited a paucity of witness support services; a lack of recompense for attending court; no court transport; and the existence of the automatic right of appeal (by virtue of s.108 of the Magistrates' Courts Act 1980, appeal is by way of full rehearing as per s.79(3) of the Supreme Court Act 1981), which meant that witnesses may have to attend the initial application and then an appeal hearing. One solicitor described their experience of resident witnesses in ASBO applications thus:

> It was obviously a worrying time for them. A number of them had to take unpaid leave to attend the initial application and then appeal hearings. It was impossible to explain [to the witnesses] that the defendants had an automatic right of appeal and that they would need to go through the horrendous experience again – especially when the defence barrister was overly aggressive in his cross-examination. In their position, I would not have agreed to be a witness!

While the majority of solicitors nonetheless supported the use of civil rules of evidence in ASBO applications, nearly half of respondents were dissatisfied with the operation of ASBO court proceedings. Fifty-nine (43 per cent) survey respondents described the 'urgent' need for the introduction of case management powers for ASBO applications. A lack of inter-agency consultation and co-operation; inconsistent attitudes towards information sharing; the presence of inexperienced evidence gatherers; the defence rarely serving evidence before trial; vague hearing dates; and, a disjointed framework for the ASBO process with different procedures in different courts, were all contributing factors that respondents argued necessitated the creation of powers to enable the courts to apply rigorous case management to ASBO proceedings.

Additionally, nearly half of solicitors (46 per cent) detailed the continuing difficulties that they were encountering with the *speed* at which the court deals with listing ASBO applications; and also with obtaining early court dates in urgent interim order cases. This has implications for those cases in which victim intimidation is a factor. Although it is 'good [court] practice' to list the first hearing of an application quickly so as to ascertain whether it can be contested, and if so, to identify the issues in the case (JSB, 2007), a fifth of respondents stated that the approximate average length of time an ASBO case was taking to come before the court, from Summons to Final Hearing, was more than 19 weeks.[14] By way of illustration, one solicitor stated that when an application is contested, the hearing will not take place for between six and nine months.[15] Thus, with regard to the use of civil procedure in ASBO applications, there are salient questions arising relating to due process and the use of evidence. In particular, if witness intimidation necessitates the use of hearsay and professional witness evidence, what protections remain – or are required to be put in place – to safeguard the rights of defendants? How are the rights of the defendant being reconciled with the rights of victims of anti-social behaviour – and with the interests of the wider community? These questions will be considered more fully in the second part of this section. First, however, we must consider the nature of ASBO prohibitions themselves, and the way in which they are currently being formatted during the court process.

Prohibitions

'Prohibitions' are the conditions that are contained within an ASBO and detail conduct which is prohibited under the terms of the order. These prohibitions can also detail geographical areas that the ASBO holder is not permitted to enter and may, for example, include specific curfews and/or prohibitions against carrying or being in possession of particular items. The principle that each individual prohibition must be 'necessary' was introduced by s.1(6) of the Crime and Disorder Act 1998. In England and Wales, the leading case on ASBO prohibitions is *R v Boness [2005] EWCA Crim 2395*, which states that any prohibition imposed must be *necessary* for the purpose of protecting persons from further anti-social acts by the defendant, and *proportionate* to the anti-social behaviour in question. Similarly, in Scotland, statutory guidance on ASBOs states that:

> [T]he terms must be only those necessary to protect persons in the area of the local authority from further anti-social acts or conduct. They can be prohibitory only. ... They should be specific, and in terms that are easily understood so that it will be readily apparent to the person and to the local community what constitutes a breach. (2004: para. 109)

It is important to note that the Crime and Disorder Act 1998 does not set any limit on the number of prohibitions that may be included in an order – nor is the Act prescriptive in terms of the type of prohibitions that may be imposed. It is only necessary that prohibitions are negative in nature, and that they are necessary to prevent further anti-social acts by the defendant. Although it is good practice that applicant authorities should provide a draft of the prohibitions sought (and standard lists should not be used) – it is for the court to decide the final wording of the prohibitions of an order.

In *R v P (Shane Tony) [2004] EWCA Crim 287, R v McGrath [2005]* and *W v DPP [2005] EWCA Civ 1333*, the principles that apply to the construction of prohibitions in England and Wales were further developed. It emerged from these cases that prohibitions should be *specific* in nature, and apropos to the individual defendant's circumstances; that they should be concise and formatted in terms that are clear and not confusing to the defendant; that prohibitions ought to be *prohibitory only* in nature (and not mandatory); that prohibitions should be constructed with reference to what is *proportionate* to the harm to be guarded against; and that the prohibited act(s) need not of themselves give rise to harassment, alarm or distress (JSB, 2007). Thus we can see that the courts have developed a relatively detailed list of principles underpinning the formation of prohibitions.

Yet, one of the other most fundamental criticisms of the ASBO model has been that the prohibitions contained within the orders are often disproportionate and unduly onerous. Local and national media have often reported the most 'extreme' (or perhaps, farcical) terms of ASBOs: for example, one ASBO prohibited a woman from feeding gulls in her back garden, and another order was given to a person preventing them from gardening in their underwear. While such examples are unique in terms of the behaviour being prohibited, it is certainly common for broad *geographical* restrictions to form part of an order's prohibitions. Simester and von Hirsch (2006: 188) contend that broad geographical exclusion conditions are 'a significant deprivation' which are sufficiently onerous in terms of the restrictions placed upon the perpetrator, to require sufficient 'fairness constraints' in the form of substantive procedural protections. Indeed, there is evidence that ASBO prohibitions can be formatted very broadly in terms of their application, and, in some cases, with a lengthy or indefinite duration attached to the conditions. If we now examine the issues arising from how prohibitions are formatted in the Scottish courts, and then in the Courts in England and Wales, we can see that there are indeed a number of salient issues arising which must be considered further.

In the Study, judicial responses in the Scottish courts to questions about the nature of ASBO prohibitions generally fell in to one of two categories: (1) those Sheriffs who that stated that they generally found the prohibitions drafted by applicant authorities to be 'proportionate' and 'reasonable', and (2) those Sheriffs who cited significant concerns as to the conditions that were being sought by some agencies. For example, one Sheriff described his refusal to grant badly drafted orders thus:

> [T]hey are very badly drafted...which means that certainly when they come before me...they have a hard time getting them through. And sometimes they have been refused simply because they are so badly drafted and they are sent away to draft them properly. I have to say that when they come back in another form it's almost as bad as the first attempt.

In particular, those Sheriffs who stated that they often found ASBO prohibitions to be badly drafted cited broad geographical restrictions as being *the most common problem encountered* in this context. Similarly, a Sheriff in another jurisdiction felt very strongly that the prohibitions that were being sought by the local authority were *not proportional*. Interestingly, several Sheriffs who had said that they were generally satisfied with the prohibitions put before them stated that when ASBOs had first been introduced, they had experienced some difficulties with applicant authorities seeking disproportionate prohibitions. However, these Sheriffs stated that such difficulties were simply early complications, or 'teething problems', and that the local authorities and the solicitors had since learnt from the refusals to grant these types of prohibitions, and were now competent in drafting orders that would meet with the standards required by the court. Alternatively, Sheriffs who were dissatisfied with the drafting of the prohibitions of the orders felt that both applicant authorities, and solicitors, were failing to learn from past mistakes, and moreover, that they had not yet developed a rigorous and effective method of formatting ASBO prohibitions. For example, a Sheriff described the complacency of solicitors in constructing the terms of the orders thus:

> What I find the council are trying to do [with ASBOs]... they are fundamentally intellectually lazy about them – that's the solicitors – they are intellectually lazy about them. They don't treat them like a conveyancing document which is what they should do. A formal contract – they don't treat it like that. And I suspect that whoever is instructing them, presumably the police, haven't really worked out themselves what they really think the danger is, to produce these blanket things for areas. I'm far from convinced they're effective. They don't allow for the obvious things that people have to do in the course of a day – you know, banned from certain areas from 7pm til 7am – you may have no choice about being in the area.

It became apparent in the course of the interviews that those Sheriffs who found the conditions of the orders to be 'necessary' and broadly 'proportionate' were those Sheriffs who described having a *good working relationship* with the local council and with the solicitors involved in the ASBO process. By way of illustration, those Sheriffs repeatedly used phrases such as 'those who apply to us from our local authority are very responsible and able people', 'the local authority do good work here', and 'we have a good bar here – and we listen to them, we know that they are not going to mislead us'. It was also evident that the *quality of the relationship* between the Sheriff and the local authority personnel involved in ASBO applications played a crucial role in the outcome of ASBO actions with regard to circumstances involving potential mitigating factors (such as addiction and mental health problems).

When Sheriffs were asked to determine whether the prohibitions of ASBOs that they had presided over more often related to behaviour that could be said to be preparatory to the commission of a criminal offence, or whether prohibitions more frequently related to behaviour that was already criminal, the majority responded that prohibitions contained a mixture of both types of behaviour. Scottish statutory guidance (2004: para 19) on the use of ASBOs states that: 'ASBOs for adults are intended to tackle behaviour that is likely to escalate to the criminal level, and patterns of behaviour which cumulatively cause considerable alarm or distress to the community. An ASBO is not intended as a substitute for criminal proceedings where these are appropriate.' However, the majority of Sheriffs stated that they were *very uneasy* about criminal behaviour forming the basis of ASBO prohibitions. Of the Sheriffs who expressed this concern, all were of the view that inculcating prohibitions on criminal behaviour was 'inappropriate' and/or 'ill-conceived'. For example, one Sheriff described the fundamental difficulty in prohibiting criminal behaviour thus:

> [E]ssentially I think that criminal conduct should be the preserve of the police and this strikes at the very heart of ASBOs – the effect of them is to render criminal, conduct which would not otherwise be criminal, and I do have slight reservations about that to be honest, and I think a lot of authorities do.

In particular, Sheriffs noted the difficulties for local authorities where an expectation existed among residents in certain locales that the council should be the 'primary agency' involved in addressing community problems such as drug dealing and aggressive behaviour. The majority of Sheriffs stated that they believed that criminal behaviour should remain a matter for the police – and not the local/housing authority. Although it was observed that prohibitions on criminal behaviour could be a useful and effective means of avoiding the criminal process for more minor infringements of the law, the majority of Sheriffs were of the view that any criminal behaviour that was violent in nature should not be a matter to be addressed by the council or housing authority. Moreover, in terms of the collection of evidence, several Sheriffs stated that they did not believe that it was appropriate for council and/or housing officials to gather evidence in circumstances where there might be a risk to their personal safety – notably in those circumstances where violent/aggressive behaviour was a feature of the ASBO application. One sentencer argued that: 'The police should be there. I mean, how do they get the particular evidence to support their application? Are they going to send people out to observe what these lads are doing? It's a nonsense! It's a police job'.[16]

Similarly, it is evident that there has also been a level of confusion within the English courts about the use of prohibitions for criminal behaviour. In *Boness*, the Court of Appeal signaled that the use of prohibitions to proscribe behaviour already deemed criminal within the law, does not automatically address the fundamental purpose of the ASBO – which is to act as a preventative (rather than a criminally punitive) order. Their lordships indicated that it is *preferable* for the courts to make an 'anticipatory' form of order with the aim of preventing the anti-social behaviour from taking place. For example, where the identified anti-social behaviour is fly-posting, the court may make an order preventing a defendant from being in possession of posters, paste and/or any other such materials designed for the purposes of attaching publicity material to buildings. Yet, in the case of *R v Barnard [2006] EWCA Crim 2041*, the Court of Appeal held that prohibitions which prevented the defendant (who had been convicted of theft from a motor vehicle and attempting to take a vehicle without consent) from touching any unattended vehicle without permission or having any object for breaking glass in his possession, would not be useful in providing any additional protection from the identified anti-social behaviour than already existed in the police power of arrest and subsequent prosecution under the criminal law. Alternatively, the court in *Hill v Chief Constable of Essex [2006] EWCA Crim 2041* allowed an order containing prohibitions which prevented the defendant from carrying a knife or bladed article in a public place. In *Gillbard v Caradon District Council [2006] EWHC Admin 2633*, Waller LJ went further by stating that an order could legitimately include prohibitions prohibiting criminal acts, and that any bar on the use of prohibitions in this way would necessarily 'drive a coach and horses through [the Act]'.

As I have argued elsewhere (2007: 428), the judiciary is a 'primary definer' on ASBOs. That is to say that

> ...through the interpretation of subjective legislative terminology and the use of their discretionary powers, judges are able to make pivotal jurisprudential decisions on ASBOs relating to civil liberties and the wider issues of law and order and crime control. While it is the local authorities and the police that are instructive in determining ASBO *applications*, it is the judiciary that primarily defines their legitimacy, their purpose and scope, and their function in law.

Hence, while the courts' approach to the formation of prohibitions has to date been somewhat incongruous (both in England and Wales, and in Scotland), their approach is likely to become more uniform as both applicant authorities, and the courts, have greater experience of the ASBO process. For example, evidence from the courts in Scotland shows the judiciary's reluctance to grant orders where prohibitions were badly drafted and also demonstrates the development and progression of judicial rationale and legal protocol(s) which are becoming more standardised.

Breach

The court's quantification and/or proportionality assessments relating to ASBO breach proceedings also has implications for due process, the rule of law, and its application.

Figures on ASBO breach rate for England and Wales to the end of 2005, show that 47 per cent of ASBOs granted had been breached (Home Office, 2006b), while a more recent study by the National Audit Office (2006) found that, of the cases studied, 55 per cent of those with ASBOs had breached their conditions. In Scotland, a total of 544 ASBOs (interim and full) were reportedly in force as at 31 March 2005. Of these, 140 (26 per cent) were allegedly breached during 2004/05 (Scottish Executive, 2005a, b). In all jurisdictions breach must be proved to the criminal standard of proof. The burden of proof remains with the prosecution, and it is for the prosecuting solicitor to prove a lack of reasonable excuse if this is raised as a defence to the action.

It was held in the English case of *Parker v DPP* [2005] EWHC 1485 (Admin), that the severity of a breach should be determined by a consideration of the individual and specific facts of a case, to include; the nature of the conduct, how soon the order was breached after it was made, and whether there was a repetition of the same breach. That is to say, each case must turn on its own facts. While the Judicial Studies Board (England and Wales) has made clear that breaches are to be treated as 'a serious matter...A court should be wary of treating the breach of an ASBO as just another minor offence...An ASBO will only be seen to be effective if breaches of it are taken seriously' (2007: 28) – it also distinguishes breaches which do not involve harassment, alarm or distress. In such cases, it is suggested that community penalties should be considered by the court as an alternative to custody, in order to 'help the offender to live within the terms of the ASBO' (ibid.). Where a community penalty is not available, it is stated that the custodial sentence should then be kept to a minimum. The Home Secretary Alan Johnson has recently said that he wants to ensure that 'every breach of an ASBO' is prosecuted by police. It will be interesting to note if this subsequently impacts upon the number of breach proceedings coming before the courts.

Macdonald (2006: 792) states that: 'One of the primary objectives of the ASBO was...to provide a mechanism for the imposition of composite sentences on perpetrators of such behaviour, that is, sentences which reflect the aggregate impact of a course of conduct as opposed to the seriousness of a single criminal act.' However, he also observes that courts sentencing defendants for breach of an ASBO have failed to impose composite sentences, citing *McCann* as the most likely reason for this (p. 795). As previously discussed, the court's ruling in *McCann* (that ASBO proceedings were to be classified as civil) subsequently meant that findings of fact from proceedings for the imposition of an order could not later be employed at proceedings for breach – a principle which was emphasised by Lord Steyn's statement that ASBO proceedings are 'separate and independent' from proceedings for prosecution for breach of an order ([2002] UKHL at 23). As a result, Macdonald has argued (2006: 796) that the body of case law on the sentencing tariffs available for breach is confusing and inconsistent.

However, the decision of the Court of Appeal in *R v H, Stevens and Lovegrove* [2006] EWCA Crim 255 has since set a precedent that breach of ASBO conditions (in England and Wales) should be treated as a *distinct* offence in its own right – undermining the outcome of the earlier case of *Morrison* [2006] 1 Cr. App. R. (S) 488 (85) which had found that the sentence for breach should be limited to the statutory maximum for the criminal offence. Their lordships' decision in *Lovegrove* thus entitles the court to impose a sentence *greater* than that given for an identical offence. Similarly, in *R v Lamb* [2006] 2 Cr App R(S) 11, the Court held that the approach previously adopted in *Morrison* could necessarily lead to the substantive impact of anti-social behaviour on the public being overlooked. However, the matter is by no means definitively settled: it has not yet been established by how much a sentence for breach can be greater.

Hence, while a precedent exists for sentencing in breach proceedings in England and Wales, appropriate sentencing tariffs are still not clear – it has been argued that some sentences imposed appear to be disproportionate to the harm caused. For example, in the aforementioned case of *Stevens and Lovegrove*, Stevens, an alcoholic in his mid-fifties, received eight months imprisonment for swearing and nine months imprisonment for being publically intoxicated. Moreover, an examination of relevant court decisions on breach proceedings in England and Wales (see, amongst others, *R v Boness* [2005] EWCA Crim 2395, *W v DPP* [2005] EWHC Admin 1333, and *R v Kirby* [2005] EWCA Crim 1228) certainly demonstrates that confusion exists within the courts as to available tariffs, the principle of composite sentencing, and, its relevance to the ASBO model.

As a result, the Sentencing Advisory Panel has recently produced a consultation document on sentencing for breach of an ASBO. Citing their frustration at the lack of sentencing guidelines in this area coupled with the absence of any publicly available data on sentencing for breach, the Panel proposes that sentencing in breach proceedings be based upon the relationship between the varying levels of harm that breach may involve and the offender's culpability in breaching the terms of the order. Additionally, the document considers to what extent the original conduct (which led to the imposition of the ASBO) is relevant in breach proceedings; mitigating and aggravating factors are identified; sentencing starting points and ranges are proposed for adults; and the principles that should apply for sentencing in breach proceedings where the offender is under 18 years of age are also suggested. The Panel proposes the following guidelines:

> 75. In line with the Court of Appeal decision in *Lamb*, the proposals are linked to whether a breach involved actual, intended or foreseeable harassment, alarm or distress. The Panel is of the view that breach of an ASBO will normally be serious enough to warrant the imposition of a community order. When imposing a community order, the court must ensure that the requirements imposed are proportionate to the seriousness of the breach, compatible with each other, and also with the prohibitions of the ASBO if the latter is to remain in force.
>
> 76. The Panel takes the view that the custodial threshold normally will be crossed where the breach involved harassment, alarm or distress. Even where the threshold is crossed, a custodial sentence will not be inevitable. The Panel also takes the view that the custodial threshold may be crossed where there has been a series of breaches where no harm was caused.
>
> 77. Where a series of breaches has involved serious harassment, alarm or distress, the Panel considers that a higher starting point will be required and, in the very small number of most serious cases, a sentence beyond the range will be justified. (Sentencing Guidelines Council, 2007: 27–8)

The table 6.1 provides details of the proposed sentencing ranges.

Additionally, the Panel suggests the following aggravating/mitigating factors to be considered in sentencing for breach (Table 6.2).

With regard to the sentencing of persons under the age of 18, the Panel proposes that in the majority of cases, the appropriate sentencing tariff will be limited to a community order, and as a result the custody threshold should be set higher than for an adult offender. In

Table 6.1   Sentencing ranges (Sentencing Guidelines Council)

Type of activity	Sentencing range
Custodial sentence Breach involving serious harassment, alarm or distress or Series of breaches involving lesser degree of harassment, alarm or distress	Starting point – 26 weeks imprisonment Range – Custody threshold to 2 years imprisonment
Breach involving lesser degree of harassment, alarm or distress or Series of breaches involving no harassment, alarm or distress	Starting point – 6 weeks imprisonment Range – Community Order (MEDIUM) to 26 weeks imprisonment
Non-custodial sentence Breach involving no harassment, alarm or distress	Starting point – Community Order (LOW) Range – Fine Band B to Community Order (MEDIUM)

Source: (ibid: 29)

Table 6.2   Aggravating and mitigating factors

Additional aggravating factors	Additional mitigating factors
1. Breach of more than one prohibition of the order.	1. Breach occurred after a long period of compliance.
2. Offender has a history of disobedience to court orders.	2. Breach was of the least significant of a range of prohibitions.
3. Breach was committed immediately or shortly after the order was made.	3. Harm caused was not foreseeable.
4. Breach was committed subsequent to earlier breach proceedings.	
5. Use of violence, threats or intimidation.	

Source: (ibid.)

cases involving serious harassment, alarm or distress, or in cases where a series of breaches have taken place (involving a lesser degree of harassment, alarm or distress) then the threshold will generally be identified as having been crossed. In the instances where the court deems a custodial sentence to be unavoidable, then the sentencing range proposed begins at four months detention, rising to 12 months detention for the most serious cases (2007: 35).

The Panel's proposals are useful in providing a framework to consider the overarching principles to be applied in formatting individual sentencing tariffs. However, as the Sentencing Advisory Panel's consultation paper identifies, there is currently a paucity of data available on sentencing for breach and so a critical aspect of current sentencing practices was not discussed within the document – *judicial rationale* influencing decision making in individual breach proceedings in the lower courts. The research findings presented below, *although limited to cases in Scotland*, provide some interesting new data on sentencing and the rationale behind sentencing practices in breach proceedings which, as we shall see, have a wider relevance to cases south of the border.

In Scotland, very little case law or research evidence exists on breach proceedings in ASBO cases, with no definitive legal precedent on the maximum penalty available when the breach involves a criminal offence. Data collected on behalf of the Scottish Executive (2005a) suggests that the term 'breach' is not consistently understood by applicant authorities, and moreover, that methods of statistical data collection within local authorities relating to types of breach are patchy and inconsistent. Local authorities and RSLs display a variance in the interpretation of statutory terminology and in the recording and collating of data on breaches in ASBO cases. Nonetheless, statistics for the period 2004/05 show that the majority of alleged breaches in Scotland were reported as having resulted in further court action. Just over a half of alleged breaches were reported to the Procurator Fiscal and a further 23 per cent involved the perpetrator being detained in custody for an appearance in court. In 14 per cent of cases was no action taken following initial police or officer visit (Scottish Executive, 2005).

Given the broad nature of ASBO prohibitions and the existence of a level of dissatisfaction among some Sheriffs in Scotland that orders were often poorly drafted (see above, *Prohibitions*); it is perhaps not unexpected that several Sheriffs interviewed for the Study said that they felt reluctant to take seriously 'minor' breaches of ASBO prohibitions, such as entry into an exclusion zone (with no accompanying anti-social behaviour). One Sheriff explained their view on the technical breach of conditions thus:

> I'm not one who goes in for standing on the ceremony of the Court.
> I'm not a great one for punishing people for flouting a court order
> or ignoring the authority of the Court. I need to be persuaded that

there is some substance to the complaint. It is sometimes a constant battle – you frequently come across it at bail application, it happens nearly every day, the Court might impose, for example, a curfew condition and you might have somebody whose committed a technical breach by being five minutes later than they should have been and he'll be arrested by the police and be charged with breach of his curfew, and because it's a breach of a court order, the Crown will take the view that this should result in the refusal of bail. They will hotly oppose bail on the ground that the individual is demonstrating a disregard for court orders, virtually by reason of the nature of the offence, because it's a court order, they ask the Court to oppose bail. And obviously, as I say, on a daily basis I have to consider debates about that. So that's just really to illustrate the point that I don't believe in punishing people just for the technical breach of court orders – there are so many circumstances that can lead to that, and for that reason I think that it would be a mistake to adopt that sort of approach in relation to anti-social behaviour orders.

Several Sheriffs expressed the view that they supported the use of ASBOs as a means to avoid the criminal process (if appropriate), in so far as they believed that prohibitive orders could potentially act as a diversion from the criminal process and the 'filling up of jails' with individuals who had committed relatively minor acts of anti-social behaviour. However, there was also a concern that punishing minor/technical breaches of prohibitions could undermine the use of ASBOs and the potential for them to be used as an effective means for addressing problematic behaviour(s) without necessitating the criminal process. While they acknowledged that prosecution was appropriate for specific types of breach involving alarm and distress, they took the view that 'technical' breaches were often innocuous enough that they ought not to be brought before the Court.

Additionally, Sheriffs' opinions varied widely as to what extent the court has regard to the maximum sentence for a criminal offence in the sentencing for breach, and, moreover, what role the ASBO plays in providing an increment in sentencing for persistent acts of anti-social behaviour. Sheriffs' responses to questions about sentencing procedures and decision-making on breach were thus determined by whether (1) they took the view that the primary function of the ASBO was to allow increased penalties for behaviour which was a culmination of anti-social acts, or (2) they were of the opinion that in circumstances involving criminal behaviour, breach should not be afforded a different or elevated legal standing in proceedings.

Those Sheriffs who took the view that the primary function of the ASBO was to allow increased penalties for behaviour which was a culmination of anti-social acts, decided sentencing for breach accordingly:

> I think the ASBO is there for a purpose – to augment the available penalty. I wouldn't feel restricted to the penalty for the offence itself. I tend to treat it in much the same way as a bail aggravation, and put on an extra month. I mean if it's a breach of two or three bail orders, as it sometimes is, I'll put on a month for each one.

In contrast, the other Sheriffs were of the opinion that in these circumstances, breach should not be afforded a different or elevated legal standing in proceedings:

> I wouldn't have any regard to [the maximum sentence for the offence in sentencing for breach] to be honest, I would just consider it on its merits. But it would be bound to be coloured by my subconscious views as to what's an appropriate sentence for the crime in the end. But I wouldn't give a breach of an ASBO some special status.

Similarly, one Sheriff explained that the civil law nature of the ASBO as a preventative measure, as opposed to a punitive sanction, influenced their approach to sentencing for breach, in the respect that this particular Sheriff was of the view that criminal behaviour should necessarily be prosecuted in the criminal courts:

> I would be unlikely to exceed the statutory maximum [for the offence], because in these circumstances I would have expected the matter to be reported to the police and then to proceed by way of prosecution as opposed to by ASBO. An ASBO is a way that I see of trying to prevent the need for people to be prosecuted and therefore liable to a criminal sanction at an earlier stage. And with a lot of people it works.

Sheriffs were also aware of the potential for a 'twin-track approach' to the sentencing of similar (or near identical) criminal acts in ASBO breach proceedings:

> [T]he penalty for breach of an ASBO could far out strip the penalties for the original crime…and I think that there's an example of that

happening in a case in England. Because the judge took the view that 'this is a court order now', it's breach of a court order – that is more serious than the original thing you were doing before the ASBO. Well, I think that there is some merit in that approach. I can see why he comes to that view but the danger is that you end up – if you had just prosecuted it properly, you would have had such and such a penalty, but because its become this sacred court order never to be breached, then you end up with far more. But that's not my experience.

The variation in the opinions of the Sheriffs on sentencing for breach, and in particular, the views of those Sheriffs who were of the opinion that the purpose of the ASBO was to augment the available penalty for a criminal act, means that – derivatively – different penalties apply for criminal acts, dependent on whether they have status within a court order. However, as several Sheriffs observed, a court order (in the form of an ASBO) will often be the result of a culmination of persistent acts of anti-social behaviour – so in their view it was wholly legitimate that breach could be treated more seriously than an individual criminal act. With regard to ASBOs creating a 'twin-track approach' to identical criminal acts, public confidence in the sentencing process was largely seen as being irrelevant by most Sheriffs. Even those Sheriffs who disagreed with a 'twin-track approach' to sentencing, did not believe that public confidence should be a factor for consideration: 'I can understand that public confidence might be affected but I think that judges as a whole are not particularly willing to take into account public opinion which is frequently uninformed'.

However, the way in which ASBO breach proceedings are decided goes to the heart of debates about sentencing, punishment, and equality before the law. How sentences are decided for those convicted of breaching their order determines important questions about the role of evidence and, moreover, notions of proportionality. Indeed, Simester and von Hirsch (2006: 189) argue that 'punishment ought, in principle, to be determined primarily by reference to the gravity and wrongfulness of [the perpetrator's] actual conduct ... not by reference to the fact per se that it has been prohibited by the order.' Indeed, these concerns are specifically pertinent to the use of interim orders as we shall now see.

Interim orders

Interim orders are available under s.1D of the 1998 Act (as amended by s.65 of the Police Reform Act 2002) and s.7 of the 2004 Act in Scotland (as amended by s.44 of the Criminal Justice (Scotland) Act 2003). This temporary order can impose the same prohibitions and has the *same penalties* as breach of a full ASBO.[17] An interim order can be made at an initial court hearing held in advance of the full hearing if the court is satisfied that the specified person has engaged in anti-social behaviour and that an interim order is necessary for the purpose of protecting the public from further anti-social behaviour. As such, there is no explicit legal provision for any representations to be made by or on behalf of the defendant before an interim ASBO is granted. In Scotland, if the initial writ has been served for an interim order, the Sheriff may dispense with intimation of the motion for the interim ASBO and grant it without hearing the defender, although the Court can consider any such representations as it sees fit. The Sheriff may grant an interim order provided the individual named on the application has received intimation of the initial writ and the Sheriff is satisfied that the anti-social conduct complained of would be established when a full hearing takes place.

In England and Wales, an interim order can, with leave of the Justices' Clerk, be made *ex parte* (without notice of proceedings being given to the defendant). In *Kenny v Leeds Magistrates' Court [2004] EWCA Civ 312*, the Court of Appeal held that an interim order made without notice to the defendant did not contravene Article 6 of the European Convention on Human Rights. Statutory guidance on interim ASBOs suggests that applications will be appropriate, for example, where the applicant authority believes that persons need to be protected from the threat of further anti-social acts which might occur before the main application can be determined. In England and Wales, where an interim order is made *ex parte*, good practice guidance states that the court should arrange an early return date. An individual who is subject to an interim order then has the opportunity to respond to the case at the hearing for the full order, and may also apply to the court to have the interim order varied or discharged. Moreover, in *R (Manchester City Council) v Manchester City Magistrates' Court* [2005] EWHC 253 (Admin) it was held, by the Divisional Court, that the Justices' Clerk should have regard to a variety of factors (not limited to) the likely response of the defendant on receiving notice of the complaint; the gravity of the alleged behaviour; the nature of the prohibitions sought; and the rights of the defendant. The 1998 Act does not, however, give any indication as to whether or not evidence has to be heard (even in part) or whether or not the interim matter can be based on representations only. Although, it was observed in the English case of *R (Keating) v*

Knowsley Metropolitan Borough Council [2004] EWHC 1933 (Admin), that, where a court is concerned with interim proceedings, it must bear in mind that no findings of fact have been made, any allegations have not been proved, and the defendant has had no opportunity to challenge the allegations.

In the course of the Study, Sheriffs in Scotland observed that (although there is no explicit legal provision for any representations to be made by or on behalf of the defendant before an interim ASBO is granted) there were instances where evidence was produced at an interim stage (such as productions lodged, convictions referred to, plans/details of property location(s) used, witness statements presented, and so on). Generally, Sheriffs' observations on interim order proceedings feel into one of two different categories: (1) those Sheriffs who had found interim orders were being used effectively and appropriately, and (2) those Sheriffs who were very concerned by the *prosecution* of interim order breaches. Those Sheriffs who were, for the most part, positive about the use of interim orders observed that they had been effective in preventing anti-social behaviour: 'My experience of them here is that one wonders why [the council] are taking so long to apply to the court!'

However, the Sheriffs in the second category were deeply concerned by the prosecution of interim order breaches, the validity of which had never been tested in court by the hearing of evidence. Those Sheriffs took the view that this 'unsatisfactory' and 'ill-conceived' aspect of the legislation was open to abuse by applicant and other agencies involved in the ASBO process. For example, one Sheriff stated that, although they were granting interim orders, they were uncomfortable about doing so:

> I must say, I doubt the validity of antisocial behaviour orders being granted without evidence. I feel uneasy about interim orders – I grant them because the legislation says that I should if I am satisfied on the basis of information given to me – but I might one day refuse and see if the council appeal me, because I think the matter needs to be looked at.

Moreover, several Sheriffs questioned the premise upon which interim orders were based. It was suggested that interim orders were, in some cases, being used to 'get round' existing legal barriers to prosecution, and were seen by some agencies as a means to avoid traditional encountered difficulties – and safeguards – in the legal process. One Sheriff, while acknowledging that (they believed that) criminal sanctions for interim order breaches were necessary, also noted that there existed important questions surrounding the legal nature and purpose of the interim ASBO:

> I think if you allow interim ASBOs to be granted at all, if you provide for them in the Act, then you're going to have to have a penalty for breach – which means a criminal penalty. The question to my mind is a more fundamental one, as to whether they are a means of getting round the difficult job of actually prosecuting somebody. But I think that if you have them, it's inevitable that you need a criminal penalty for breach. (S6)

It was evident that a significant proportion of Sheriffs interviewed felt that the prosecution of interim order breaches was a matter that 'needed to be addressed'. Indeed, the administration of interim orders – and the statutory provisions governing their use – raises several important questions specifically in respect of procedural fairness. First, because there is no legal requirement that evidence should be led at the interim stage, this necessarily means that interim orders can be issued without the lodging of any productions, or the hearing of any witness statements. As a result, interim order breaches can be prosecuted in court when the validity of the original order had never been tested by evidence at the initial hearing. The implications of the use of interim orders granted *ex parte* (and the prosecution of any breaches arising) for the safeguards associated with due process will be examined in detail in the second part of this section.

Orders on conviction (CrASBOs)

Following legislative changes made in s.64 of the Police Reform Act 2002 and s.234AA of the Criminal Procedure (Scotland) Act 1995, criminal courts may now also make orders against individuals convicted of a criminal offence. In a similar way to ASBOs imposed in the civil courts, ASBOs on conviction are intended to prevent further anti-social behaviour, but specifically in relation to incidents that the police have reported (and where criminal proceedings have subsequently been taken). An order on conviction is granted on the basis of the evidence presented to the court during the criminal proceedings and any additional evidence provided to the court after the verdict. Contrary to reports by Madge (2004) that orders on conviction are often regarded as a component of a sentence, the order on conviction is *not* part of the sentence and can only be made in addition to a sentence or a conditional discharge.

While the ASBO on conviction was never intended as a replacement for orders on application, they *were* intended as a means of expediting a lethargic and resource-intensive court process. In England and Wales, the amount of orders now obtained on conviction exceeds the volume of orders obtained by section one stand-alone applications. For the period between April 1999 and September 2004, of those ASBOs issued in England and Wales, 59 per cent were on application and 41 per cent were on conviction (House of Commons, 2005b). Yet, orders on conviction only became available to persons in England and Wales who had been convicted of a relevant offence committed *on or after* 2 December 2002, while ASBOs had been available since 1 April 1999. Statistics for England and Wales from November 2002 to September 2004, show that the number of orders granted on conviction accounted for 71 per cent of all ASBOs issued in England and Wales during this period (Burney, 2005: 94).

There are no court rules setting out the procedure to be followed in applying for an order on conviction. The Court of Appeal has, however, provided some instruction in this area in the case of *R v W and F [2006] EWCA Crim 686*. Having noted the absence of court rules setting out the procedure to be followed in such cases, the Court of Appeal gave the following general guidance:

- The prosecution should identify specific facts said to constitute anti-social behaviour;
- If the defendant accepts those facts, then they should be put in writing;
- If the defendant does not accept them, they must then be proved to the criminal standard of proof;
- The defendant should have sufficient time to consider the prosecution's evidence against him/her – particularly with regard to evidence that is beyond the scope of the offence that the defendant has been convicted of;
- Hearsay evidence is admissible;
- Procedure as per the Magistrates' Court (Hearsay Evidence in Civil Proceedings) Rules 1999 should be followed;
- Findings of the court should be recorded in writing as rule 50.4 of the Criminal Procedure Rules 2005.

As such, the conditions necessary for making an order on conviction in England and Wales are twofold. First, the court must be satisfied that the defendant has acted in an anti-social manner that is likely to cause harassment, alarm or distress. Second, it must be shown that the order is *necessary* to protect persons in any place in England and Wales[18] from further anti-social acts by the defendant.

The Court of Appeal has sought to reinforce the principle that an order should *not* be made simply for the purposes of extending the penalty for committing an offence (see *R v Kirby [2005] EWCA Crim 1228*, *R v Adam Lawson [2006] 1 Cr App. R (S) 323* and *R v Williams [2006] 1 Cr. App. R (S) 305)*. Moreover, in *R v P [2004] EWCA Crim 287*, where *P* was a habitual mobile phone thief who had received a four year custodial sentence, the court held that an order on conviction should not be granted, and that orders on conviction should *not* be made where custody has been imposed if the offender is not persistent and a period of supervision will follow. Alternatively, in *R v Scott Parkinson [2004] EWCA Crim 2757* the court stated that an order on conviction could be issued against the defendant, despite a lengthy period of custody also having been awarded. The defendant (who had been convicted of robbery) had demonstrated persistent acts of anti-social behaviour which had subsequently led to him being evicted. The court also cited the failure of other sentencing alternatives which had been tried previously without success.

However, despite the extensive use of orders on conviction in England and Wales, the courts in Scotland have made very limited use of CrASBOs. In contrast to England and Wales, where a court can make an order on conviction on its own initiative (and an application for an order is not required) or the order can be requested by the police or local authority (who may make representations to the court in support of the request), in Scotland orders on conviction are not applied for by any authority, or the procurator fiscal. Instead, it is a matter for the court based on the evidence given at trial or the Crown narration in court. Since CrASBOs became available in Scotland in 2004, only 65 orders on conviction had been granted by 2007.

No research evidence has previously existed on the use of orders on conviction in Scotland. The most recent research study on ASBOs conducted by the Scottish Executive only considered the use of interim orders and full ASBOs and did not provide any data on the use of orders on conviction in Scotland. Given the enthusiastic uptake of orders on conviction in England and Wales, and the very limited use of CrASBOs in the Scottish courts, my own research has examined the reasons behind the limited use of orders on conviction in Scotland and found that the two main reasons for the low numbers of orders in conviction in Scotland were, first, the reluctance of the judiciary to grant orders on conviction, and second, the attitude(s) of fiscals towards 'becoming involved in the sentencing process...which [goes against] one of the fundamental principles of Scots law'.

However, it was apparent that although most Sheriffs were, indeed, reluctant to grant orders on conviction, this was *not* because of their de facto opposition to orders on conviction. Instead, the reluctance was as a result of the circumstances in which orders on conviction were being sought. The majority of Sheriffs were of the view that they were very often unlikely to have been imparted with the appropriate and relevant knowledge/information (from the fiscals) that would enable them to legitimately grant such an order. Subsequently, at present, it is evident that Sheriffs are very reluctant to grant/make use of orders on conviction. As one Sheriff stated: '[Orders on conviction are] just not seen by many Sheriffs as being appropriate as a suitable disposal'. The interview findings were almost unanimous in detailing the reason(s) for this, with almost all Sheriffs stating that orders on conviction will continue to be used in a limited fashion until an appropriate protocol/system is developed with regard to the necessary information being passed to the Sheriff:

> [A]lthough the statutory power has existed for us to impose these orders, our immediate point was always: 'who is going to provide the detailed information which we need?' – not just to make the order in principle, but to do it on an effective basis. And we would need serious information, like the kind we get from local authorities and as far as I know, the prosecutors were not only not keen, but they were refusing to get involved. We've had detailed discussions about that over the last twelve months, and saying 'well yes, in principle, there's no reason why we wouldn't use that power in the appropriate circumstances' but we would have to be sure that we have an agreement where the information is going to come from. Then if the fiscal – and he seems to be the appropriate person to produce it – if he was going to do that, he'd have to depend on the police, then we'd have to give the defence a chance to object to any information. To us, we seem far away from an appropriate system whereby that could properly and effectively operate.

Most Sheriffs described the benefits that would accrue from the development of a standard protocol on information sharing, and they detailed the central importance of this aspect of the legal process in obtaining orders on conviction. It was apparent that these Sheriffs were not willing to grant orders on conviction without the requisite due process and the derivative safeguards of appropriate legal procedure(s). It was evident that the majority of Sheriffs interviewed were aware of this problem relating to the granting of orders on conviction without the necessary and requisite information, and many of those interviewed had discussed the matter with other members of the judiciary within their own jurisdiction, while others had discussed it with Sheriffs in other jurisdictions. While the majority of those Sheriffs interviewed were very positive about the prospect of the development of a proto-col on information sharing, several Sheriffs noted that there were Sheriffs in other jurisdictions who would be very unhappy about such a development:

> ...I think it would do no harm for there to be some sort of proto-col about information passing from one to the other – but I could see there being great resistance from certain Sheriffs about how this would infringe the independence of the judiciary et cetera, et cetera. But, you know, you've got to be seen to make...you've got to comply with the law and with the contention, you've got to have appropri-ate information – but who's going to give you that information? It doesn't just appear!

It was also evident that some Sheriffs were of the view that, in order for the process for orders on conviction to become successful and, ultim-ately, effective, it would be necessary for the application procedure to become longer in duration, whereby a case is continued in order for the judge to decide on matters arising from (the information contained in) the CrASBO application. However, given the already overburdened case load of the Scottish courts, it is possible that any measures introduced with the potential to bring about further delay to court proceedings may well be unpopular. Yet, it was clear that the Sheriffs (almost unani-mously) felt that the current procedure for orders on conviction was fundamentally unsatisfactory.

Hence, it is interesting to note *the reluctance of the Scottish judiciary* towards the granting of orders on conviction as a result of their dissatis-faction with the current court procedure(s) – compared with the approach of the courts in England and Wales, who have granted, despite the lack of any court rules or procedure for the making of an order on conviction,[19] a high number of orders on conviction since they became available in 2002. The courts in England and Wales have, however, sought to pro-scribe the use of orders on conviction for the purpose of *extending the penalty* for a criminal offence. In *R v Kirby [2005] EWCA Crim 1228*, the

Court of Appeal held that an order on conviction should not be made where its primary purpose was to enable the court to grant a higher sentencing tariff in the event of future offending of a similar nature (JSB, 2007: 37).

ASBOs, young people and children

The use of ASBOs for children and young people is widespread in England and Wales, with about half of all orders issued being granted against young people below the age of 18 (Home Office, 2007). Moreover, research has found that in England and Wales, very few Individual Support Orders (ISOs) have been used to support children with behavioural difficulties who have been issued with an ASBO (BIBIC, 2006). Support orders can be given to 10–17 year olds who have already been issued with an ASBO and are designed to tackle the underlying causes of the problem behaviour. However, only seven ISOs were issued between May and December 2004; compared to over 600 ASBOs issued to young people aged 10–17. Hence, there has been substantial controversy relating to the support networks that are in place to help young people with ASBOs who also have diagnosed behavioural problems. Support agencies have argued that what is needed is a tiered approach to anti-social behaviour interventions, with closer assessment of problematic behaviours in multi-agency discussions. In cases involving young people, it has been argued that youth offending teams need to be involved at an early stage of a young person's problematic behaviour so that learning or behavioural difficulties can be identified at the earliest stages, which would help to ensure that an inappropriate course of action is not going to be taken (YJB, 2006).

In marked contrast, the use of ASBOs for children and young people is in Scotland is limited, with only half a dozen orders having been granted to children (below the age of 16) by 2007. The Study found that Sheriffs generally agreed that it could be a difficult process to pursue orders against children, but felt that this was an important safeguard in the system:

> [The local authority] do a lot of good work to make [the use of ASBOs] unnecessary. I've met with a number of agencies which do excellent work – including with the under 16s – especially in our most troubled part of [the local authority]. Although I certainly agree that ASBOs for under 16s can be difficult to obtain. But I'm very glad that it is like that.

Several Sheriffs expressed their support for the use of ASBOs for children. However, the majority of Sheriffs were of the view that orders for children were 'ineffective' and 'irrelevant'. In particular, Sheriffs raised concerns about the possibility of young people getting drawn into the criminal justice system unnecessarily:

> I think [using ASBO against under 16s] is falling into the trap of coming down hard on the people who have been spotted...I mean I'm sure there's research that shows once an individual has come to the notice of the prosecuting authorities then the likelihood of their being prosecuted is higher, and I see it all the time here. People who offend are then given bail subject to conditions and they are very, very easily re-arrested for fairly innocuous matters. You get a crowd of youths who scatter – and the one who is caught, is the one who is recognised – and I think that antisocial behaviour orders merely put greater pressure of youngsters who are already having difficulty functioning in society and I'm not sure if that's the best way to go about it.

Sheriffs that were concerned about the use of the orders for children and young people, cited the implications upon breach, which they believed would mean that children and young people would inevitably become caught up in the cycle of children's hearings or, in the case of young people, the criminal justice system.

Some local authority practitioners in Scotland have also expressed the view that 'an ASBO should be seen as a warning [to children and young people], not a last resort'. Those Sheriffs who were generally supportive of the use of the orders for children and young people (in circumstances where such an order was genuinely deemed to be necessary) were of the view that it might, in some circumstances, be appropriate to use orders in this way. However, the majority of Sheriffs (who had expressed concerns about the use of the orders for children and young people) again stated that they did not believe that it was acceptable to use ASBOs as a 'warning' to children and young people. Of those Sheriffs that disagreed that ASBOs should be used as a warning to children and young people, most stated that alternative interventions (such as acceptable behaviour contracts (ABCs) and parenting orders) should be made much greater use of. Although these Sheriffs suggested that such interventions were used rarely, if at all: 'I think the idea of parenting orders is a great idea – but I've never seen an application for a parenting order'.

Moreover, in view of the existence of statutory measures which prevent children from being detained for breaching the prohibitions of their order, Sheriffs felt that an ASBO granted against a child or young person in Scotland would be of little consequence:

> The virtual certainty is that it goes back to the Children's Hearing, they will then either admonish the individual or they'll put them on a supervision requirement. But if they need a supervision requirement then the chances are that they are already in front of the hearing through family's and children's issues in any event, so I just think that they are of little relevance.

The issue of ASBOs being used as a 'badge of honour' by virtue of the lack of sanctions available upon breach was also raised. For the most part, the majority of Sheriffs did not believe that the use of ASBOs for children and young people, in their current legislative form, were a useful or well constructed part of anti-social behaviour legislation in Scotland, particularly in view of the existing problems associated with offending behaviour by children and young people:

> [W]e already have extreme problems about the detention of children – not anything like sufficient places, and that's only appropriate for serious criminal offending. What else are you going to do? Impose fines? That would be an absolute waste of time. So that's another reason why I regard the use of [ASBOs against under 16s] as very restricted to serious circumstances.

Similarly, Sheriffs in Scotland were almost unanimous in agreeing that publicity should not be used for ASBO cases involving children. As one Sheriff stated: 'I'm dead against it...[the use of publicity for under 16s] would be the wrong route to take entirely'. Contrary to suggestions that the use of ASBOs for under 16s is likely to 'grow in earnest' in Scotland (Walters and Woodward, 2007), the study found that the majority of the Sheriffs interviewed were not supportive of their use against children. Moreover, due to the statutory provisions in Scotland which require a level of agreement from all interested parties (social work, police, local authority) before an ASBO application for a person under the age of 16 can progress, it is a very difficult process to obtain an order against a child in Scotland. Thus, given the reluctance of the judiciary to grant orders to children, coupled with the 'onerous' statutory provisions required to proceed an application, it is suggested that the use of ASBOs for persons under the age of 16 in Scotland is unlikely to rise significantly in the near future.

Defending ASBO applications

Given the high number of successful ASBO applications (in England and Wales, and in Scotland, the Courts have refused one per cent of all ASBO applications, Home Office, 2005b; Scottish Executive, 2005) coupled with the civil rules of evidence used in ASBO court proceedings, in the interests of fairness, the automatic right of appeal is an important provision within anti-social behaviour legislation, particularly with regard to the existence of 'inappropriately issued' ASBOs. While the Home Affairs Committee on Anti-Social Behaviour has stated that 'we do not consider the inappropriate issuing of ASBOs... [to be] a major problem' (2005: 73), the Committee has also recommended that 'the Home Office commissions wide-ranging research in this area' (ibid.). Moreover, the Home Affairs Committee notes that it is 'relatively straightforward to apply to the Court... for the terms [of an order] to be varied' and that 'there is also a right of appeal' (House of Commons, 2005a: 73), it further notes that 'cases in which these options are not being taken highlight the variable quality of legal representation rather than any difficulties with the current provisions for variation and appeal' (ibid.).

However, in the Study, a number of Sheriffs stated that defence solicitors were often very reluctant to appeal the orders, even when it was apparent that there was evidently a justifiable reason for doing so. As one Sheriff argued:

> [The ASBO applications that have come before me] are badly drafted but of course the defender's solicitor is as bad in not coming to court immediately when it's been granted, you know, get back in court and argue it – not on the merits of whether it's a good idea or not – but argue it on the basis that this appallingly drafted document should not be allowed to go any further.

Other Sheriffs observed that the majority of ASBO applications in their jurisdictions were the result of a culmination of persistent anti-social behaviour which was, in most cases, essentially 'indisputable'. Hence, they felt that the solicitors saw 'no merit' in opposing such applications:

> [M]y experience is that the solicitors take the attitude that there is so much that has happened before we get to the stage of an

antisocial behaviour order that there is no point opposing the interim order.

It was also apparent that several Sheriffs were very unlikely to reject applications, even if they believed that the application was poorly constructed and/or not the most appropriate intervention, if the application was uncontested by the defence solicitor. A small number of Sheriffs felt concerned that the chaotic nature of many ASBO defendant's lifestyles (often as a result of addiction problems) might impact upon their ability to organise a defence to ASBO actions. However, the majority of Sheriffs, while acknowledging that the nature of ASBO proceedings was such that an application could succeed undefended, were for the most part, unsympathetic to the suggestion that the chaotic nature of defendants' lifestyles could mean that they were disadvantaged in court because they had not organised a defence to an application. As one Sheriff observed:

> Well, there may be [problems with people subject to an application not organising opposition], but I mean, the same people, I'm quite sure, if you said 'come along here at 12 noon tomorrow, and there'll be a party with lots of booze' – they would understand that enough.

Several other Sheriffs made similar comments and stated that the chaotic nature of many ASBO defendants' lifestyles was not a factor that they were generally sympathetic towards. However, the views of the judiciary towards potential mitigating factors such as addiction and mental health problems has important implications for the issuing of ASBOs to disadvantaged groups which we will now consider in more detail.

Disability, addiction and mental health problems

Agencies that are involved in ASBO applications are not required to demonstrate that the individual named in the application intended to cause harassment, alarm or distress – only that anti-social conduct had taken place, which has, or is likely to cause alarm or distress to others. In England and Wales, local authorities have a duty under the NHS and Community Care Act 1990 to assess any person who may be in need of community care services, which means that if there is any evidence that a person against whom an anti-social behaviour order is being sought may be suffering from, for example, learning difficulties or an autistic spectrum disorder, then the necessary support is supposed to be provided in tandem with the evidence-gathering process. However, recent research (BIBIC, 2006) has shown that relevant information regarding learning difficulties is not always made available – or is not uncovered – soon enough and potential mitigating factors are subsequently missed in court. While it is estimated that about 10 per cent of ASBOs are granted against people with learning difficulties, information on the number of ASBOs issued to people with diagnosed emotional or behavioural problems or learning difficulties is not collated by the Home Office. The reason for this is that Ministers have stated that the meaningful collation of data would be very difficult in view of the lack of a consistent and widely agreed definition of such problems. Hence, in this respect, estimates on the numbers of people with ASBOs who have learning difficulties are at present an estimate.

Moreover, because the definition of 'anti-social behaviour' contained within the legislation is so wide, it has the potential to affect people who have what are termed 'hidden' disabilities such as autistic spectrum disorders, Tourette Syndrome, ADHD and Huntington's disease as well as people with learning difficulties. As people with disabilities can display challenging, obsessive or ritualistic behaviour which may appear to some people to be anti-social, research has shown that they can be targeted by the legislation. As a result, there is evidence that individuals (including children) have been issued ASBOs who have, for example, autism, Asperger syndrome and ADHD. It has also been shown that people with learning and communication difficulties frequently experience problems in understanding the prohibitions of orders, which can relate to memory difficulties, problems with interpretation of the prohibitions, and so on.

Additionally, it is also important to note that, in the past there has been evidence of a clash between the Disability Discrimination Act and the use of anti-social behaviour legislation. For example, in *Manchester City Council v Romano [2004] EWCA Civ 834*, the Court of Appeal considered the lawfulness of the eviction of tenants for anti-social behaviour, where the behaviour concerned was disability-related. Their lordships held that possession orders in this case were justified because of the effect on the health of neighbours. However, in December 2006 the new provisions in the Disability Discrimination Act 2005 came into affect which subsequently makes it unlawful for public authorities to discriminate against people with disabilities in the exercise of their public functions. There is a view that this new legislation could help to prevent, or at least to reduce, individuals with disabilities and mental health problems being issued ASBOs inappropriately – in particular because it widens the definition of disability and introduces a new positive duty on public bodies to promote equality for disabled people. This is an aspect of the law that will require further research in due course in order to establish the extent to which it changes the use of ASBOs for people with learning difficulties.

In Scotland, the position with regard to the use of ASBOs is very similar to the approach in England and Wales and there has been significant concern that the relevant legislation and guidance does not provide adequate protection for vulnerable people, such as the disabled or those suffering from mental health problems. For example, the Scottish statutory guidance on ASBOs states:

> The authority applying for the order does not have to prove intention on the part of the defendant to cause alarm or distress. (Scottish Executive, 2004a: s.33)

The guidance goes on to note that:

> While an authority does not have to prove intention, it would not be appropriate to use an ASBO where an individual cannot understand the consequences of their actions. For example, it is highly unlikely that an ASBO would be the most appropriate means to address the behaviour of an individual with autistic spectrum disorder or any disability or other developmental or medical condition which is considered to cause their behaviour. Where an individual has such a condition, or it is suspected they may have such a condition, advice should be sought from medical experts or other bodies with expertise in the area on support which is available. (ibid: s.34)

Hence, the Scottish Executive's statutory guidance on the use of ASBOs states that it is 'highly unlikely' that an ASBO would be an appropriate measure to deal with a person whose behaviour is the result of disability. However, as disability charities have observed, this phrase has no basis in law and is open to interpretation. It relies upon a common sense approach of local authorities in interpreting the legislation, which means that essentially no legislative restrictions exist to prevent the use of the orders for those with disabilities or learning difficulties. The guidance states that in cases where an individual has a condition, or is suspected of having a condition, then advice should be sought from a 'medical expert'. Disability charities have argued that this then creates an issue of workload, of GPs being approached who don't necessarily have the appropriate knowledge or information and, moreover, that this also effectively ignores the role organisations and agencies can play in providing appropriate advice and information on behaviours that are associated with diagnosed conditions.

The study research with members of the judiciary in Scotland also found that there was evidence of a level of apprehension about how ASBOs were being administered to those with severely chaotic life-styles. Several Sheriffs raised concerns about the potential for mitigating factors such as addiction and mental health problems to be missed in ASBO proceedings. One Sheriff gave an example of a recent case as an illustration:

> In dealing with criminal matters, and particularly in sentencing, a Sheriff will not make any community based order, without first having obtained a social inquiry report at the very least. Plus a psychiatric report, plus a report from a drugs and alcohol agency, in other words, any order relating to probation, a restriction of liberty order, or any conditions attached to a probation order, or psychiatric treatment or drugs treatment or anything like that would be done after a process of investigation and advice from experts. By contrast, a council can apply to the Court, even at the interim stage, for an order which might amount to the equivalent of a community based order such as a curfew. The case of [Y] illustrated that – a prohibition against entering certain areas. And what disturbed me about that case, which is a very good illustration [of the problem], was the prohibition against possessing alcohol, and the prohibition against being under the influence of alcohol. It's only because I had an insight into the case that I knew that the young lady, who had been up before me numerous times was a very vulnerable young lady. It was only because I had that insight that I knew that she had a serious drinking problem, and she's only [A or B years old], with a serious drinking problem, so I was pretty well aware that she needs to have her alcohol problem addressed but using a court order that would render it criminal for her to have a drink, to my mind, wasn't the best way of doing it. And so I do think that there is a danger that these issues will be overlooked. If I was a visiting Sheriff, and I knew nothing about that lady, and the council came in with this litany of offending, I would have just said ok, yes, on you go. And that could have been very harmful to the young lady. So, I think that there is an issue there.

However, those Sheriffs in the smaller Courts stated that it was very unlikely that such instances would arise in their Courts because the judiciary in smaller jurisdictions are highly likely to know of the circumstances of individuals who come before them from previous cases that they have presided over:

> I think in a smaller jurisdiction like this, the chances [of mitigating factors being missed] are less because I don't think I've ever seen an ASBO application that wasn't in respect of somebody I didn't already know, and I already had quite a lot of information on them anyway, either through the criminal courts or the child and family side or whatever.

While several Sheriffs expressed concern at the potential for mitigating factors to be missed in court, the majority of remaining Sheriffs, while acknowledging that addiction and mental health problems were common features of the ASBO cases that came before them, did not think that this necessarily presented a problem for ASBO applications per se – most saw addiction and mental health problems as 'a fact of life in these type of cases' which was to be 'expected'. One Sheriff described the presence of addiction and mental health problems in ASBO cases thus:

> Yes – drink, drugs, and, either separately, or because of drink and drugs, mental health problems – it's a fact of life that the vast majority of those involved in the criminal court are going to have any one or more of these problems. But what can you do about it? Many of them have no desire to change, the facilities aren't there to assist them to change, it's a viscious circle. Until they give up the drink and drugs at the level they're taking them, their mental health isn't going to get any better – and most of them regard cannabis as being the cure for their mental health [problems] and not the cause of it!

Of those Sheriffs who saw addictions and mental health problems as being atypical of ASBO applications (but who did not believe that this necessarily presented a problem for ASBO procedure in itself), about half were sympathetic towards individuals in such situations: '[U]nfortunately, many of them are just hopeless cases...'. The other half of respondents, although not expressly sympathetic, mostly viewed the presence of these factors within a wider spectrum of criminal offending that came before the courts, in which the presence of these factors was often 'inevitable'. One sentencer observed that: '[M]yself, and all my colleagues here are very aware of ... I mean it's something like 70 per cent of our criminal convictions here are by those who are either addicted to drugs or alcohol'.

However, in the same way that it was evident that the quality of the relationship between the Sheriff and local authority personnel involved in ASBO applications played a crucial role in determining the form of ASBO prohibitions in individual applications (see above, *ASBO prohibitions*), it was also apparent that this same relationship was again highly influential with regard to mitigating factors. Sheriffs attached a high level of importance to the views of local authority practitioners, where a good relationship existed between the local authority and the Sheriff. If the individual Sheriff was of the view that the local authority professionals were competent and trustworthy, then they were much more likely to grant applications, and less likely to be concerned that mitigating factors could be missed. This was also true of the solicitors involved in the applications. It appeared that Sheriffs were more inclined to attach weight to the arguments of the bar if they, as before, believed them to be trustworthy and competent. Of course, the reverse was also true, and Sheriffs who believed that solicitors who were 'incompetent' and 'lazy' were unlikely to be 'impressed' by the ASBO applications put forward by them. It was also apparent that, given the frequent presence of the aforementioned factors in ASBO applications (addiction problems, chaotic lifestyles of defendants), a number of Sheriffs were of the view that their role in ASBO cases was as much about 'social work' as it was about deciding the law: '[T]here are times when I feel like I'm being made to be a criminal justice social worker! And I'm not! And I shouldn't be made to be'.

One Sheriff suggested that it would be useful for Sheriffs deciding on ASBO applications if a social inquiry report was provided as part of the application. Another Sheriff described the difficulty in obtaining such reports for ASBO applications thus:

> In the real world, a Sheriff [in an ASBO case] would be very fortunate if he or she can get a social inquiry report just like that – I have enough trouble getting a social inquiry report just for my criminal cases!

Overall, however, it was felt by the majority of Sheriffs that a discretionary power to request a social inquiry report might be useful in some cases, but that the power should only ever be discretionary – and not statutory.

Hence, in this chapter, the empirical data and case law presented throughout has demonstrated the significant influence of judicial decision-making and discretion on the administration, management and outcomes of ASBO use in both England and Wales, and in Scotland. Indeed, judicial discretion has been shown to be specifically influential in respect of prohibitions; interim orders; orders on conviction (CrASBOs); sentencing for breach; defending ASBO applications; and the use of ASBOs for young people and children. However, as we have observed, the way in which ASBOs are administered and the legal processes involved in their application, raises a number of important issues concerning ASBO court procedure and its (insufficient) attention to due process which will now be considered in more detail.

7
ASBOs and the Targeting of 'Undesirable' Persons

Historically, the courts in Scotland, and in England and Wales, have long expounded the general principle that both legal procedures and the decision-making of officials must be fair – and that a duty exists to act fairly and to afford all participants the right to be heard. Galligan has observed that: 'Exceptions to the principle might still be made, but only where there are strong reasons for doing so; indeed the presumption is that the general duty to follow fair procedures will apply unless exceptions can be justified. The practical application of the principle can still be uneven, with various factors influencing a court's appraisal of what procedures are needed' (1996a: 329). It will now be considered to what extent (if at all) procedural fairness in ASBO cases has featured secondary in the pursuit of policy objectives, and moreover, whether administrative and procedural decision-making in ASBO cases requires to be revised and improved. Subsequently, this chapter examines three specific areas of ASBO legal procedure which are identified as particularly problematic in the context of fairness and due process for ASBO defendants: the defence of ASBO action, interim orders granted ex parte, and the 'balancing' of interests.

Defending ASBO applications

Although ASBO applications are rejected by the courts very rarely, judicial discretion can in fact play a pivotal role in determining (and, importantly, limiting) the scope of ASBO prohibitions. However, in the context of our consideration of *fairness* in ASBO proceedings, there are two factors which are very significant for the purposes of this analysis – and that subsequently require to be considered here in more detail. These two elements are, first, the nature of the civil procedure which has been used to achieve the high success rate of ASBO applications, and second, the legislative provision for an automatic right of appeal in ASBO cases.

As we have seen, the decision of the House of Lords in *R v McCann [2002] UKHL 39* – to classify ASBO proceedings as *civil* in nature – signified that the rule against using hearsay and professional witness evidence (which applies in criminal proceedings), would not apply to ASBO applications. This judgement enables professional witness evidence and hearsay to be used in cases where witnesses are too intimidated or fearful of reprisals to give evidence themselves. Indeed, the protection of witnesses is one of the most fundamental aspects of the ASBO model. As the Study data has shown, over 90 per cent of solicitor respondents in England and Wales have continued to experience difficulties in securing witnesses in ASBO cases as a result of 'fear of reprisals'. However, while it is fair to say that hearsay and professional witness evidence was viewed by government policy-makers as, if not a panacea, then certainly an elixir, for the problems associated with obtaining evidence from fearful or intimidated witnesses, the Study suggests that the use of hearsay and professional witness evidence has not extinguished these difficulties and, as many respondents in England and Wales detailed, witnesses in ASBO cases continue to suffer intimidation and retribution (before, during and after) the court process.

Consequently, it is not surprising that interim ASBOs are almost always obtained on the basis of hearsay evidence, or, as interim order rules allow, on a *prima facie* basis, without the lodging of any witness statements or productions. However, it is perhaps surprising to discover that applicant agencies in England and Wales are able to obtain a full ASBO based only on hearsay evidence. While there are certainly legitimate grounds for the imposition of an interim ASBO based only on hearsay evidence,[1] the scope for error(s) in decision-making (bureaucratic and legal) are undoubtedly increased when the courts are willing to endorse the use of full ASBOs – with a two year minimum duration in England in Wales – obtained only on the basis of hearsay evidence. When orders are applied for and granted in this way, the opportunity for ASBOs to be used as part of a neighbour(hood)/community vendetta is significantly increased.

Yet, in spite of this, it would not be prudent to prohibit full ASBOs from being granted solely on the basis of hearsay or professional witness evidence. This would preclude the use of ASBOs, and the protection that they can provide,[2] for victims of anti-social behaviour. Rather, in view of the civil rules of evidence used in ASBO cases, and the high success rate of ASBO applications, in order to achieve fair treatment for defendants in ASBO applications, there must exist provisions which *mitigate* the conflict between the protection of individuals from anti-social behaviour, and the right of defendants to fair treatment. Within this paradigm, the provision of adequate legal representation and the right of appeal is of fundamental importance.

While the Home Affairs Committee on Anti-Social Behaviour (House of Commons, 2005a) noted the existence of the automatic right of appeal in ASBO cases, it also suggested that cases where an order had been issued 'inappropriately', and the option of the right of appeal was not taken, highlighted the variable quality of legal representation available to defendants, *rather than any difficulties with the legislative provisions for appeal*. When asked if they agreed with the opinion of the Home Affairs Committee on the variation in the quality of defence counsel, more than half of the Study respondents in England and Wales stated that they did agree. Almost a quarter of solicitors believed the quality of defence counsel available to ASBO defendants to be 'highly variable', although the majority of respondents (over 60 per cent) thought that, overall, the quality of defence counsel was of a good standard. However, with regard to the right of appeal specifically, solicitors identified the 'lack of Legal Aid' as the primary reason for low numbers of appeal in ASBO cases. Let us now consider the quality of defence counsel available to ASBO defendants, and the current limits placed upon Legal Aid.

In an adversary system, the quality of legal representation available to defendants, may, it is fair to speculate, affect the risk of erroneous deprivation of substantive rights. Given that the quality of representation depends on the ability to pay, current civil procedure doctrine would seem to provide a systematic distribution of the risk of error in favour of those who have the greatest share of social resources.

Moreover, the importance of the standard of legal representation provided to a defendant should not be understated, given the range of functions that solicitors potentially can undertake. As Galligan observes: 'Lawyers can provide advice on what must be done to gain benefits or to avoid burdens; they can help in collecting evidence and presenting the facts; they can advise on the law; and they can be especially effective in examining the facts and material upon which the deciding authority proposes to act. If the matter goes to appeal, then lawyers can provide invaluable help in assembling and presenting the case' (1996: 363). Indeed, the skills involved in legal representation should not be underestimated, since it is unlikely that defendants will be able to successfully represent themselves.

Walker et al.'s (1979) rudimentary work on procedural fairness in legal contexts found, contrary to the situation which presently predominates (where, the client, by virtue of their lesser position, both in terms of knowledge and in terms of their limited options, is *led* by their solicitor) that, in fact:

> [T]he attorney should facilitate participation by the client in the decision making process. The case ought to be regarded as belonging to the client, not to the lawyer, and the attorney should see himself as the agency through which the client exercises salutary control over the process. In this client-centred role, the attorney best functions as an officer of the court in the sense of serving the wider public interest. (1979: 1417)

However, it appears that, in the (minority of) cases where legal representation is of a poor standard (for example, where the lawyer is uninformed about legislative provisions or is disinterested in representing the client's best interests), the defendant will have few alternative options. So, in this context, there is certainly the scope for ASBOs to be issued inappropriately. In these instances, it appears that the standard of defence counsel constitutes one factor in this outcome. The government has been suggested that the high success rate for ASBO applications means that applicant agencies are applying for the orders in the correct circumstances, and are providing the requisite evidence in support of their applications. However, interview data from the judiciary in Scotland, for example, demonstrates that ASBO applications are infrequently contested, even in circumstances where there may be legitimate reasons for doing so. The potential for ASBO applications to be issued inappropriately in these circumstances would therefore appear to make the legislative provision for the automatic right of appeal *all the more necessary*. Yet, as data for England and Wales also demonstrates, the right of appeal is being taken in only a very small number of cases which is primarily a result of a lack of Legal Aid.

The current restrictions on legal aid, it has been argued, are particularly stringent. For example, in Scotland in 2005/06, an application for legal aid was made in 81 out of 344 ASBO application cases. In 43 of these cases, the pursuer (applicant authority) objected to the application. The two main reasons cited for objection were where an application was deemed 'not in the public interest' and therefore 'a waste of money' and second, where there was 'no defence to the action'. As a result, legal aid was granted in only a tenth of ASBO cases in 2005/06 (Scottish

Executive, 2005a: s.4.4). This figure is the same as the previous year. Similarly, in England and Wales, the reforms to Legal Aid resulting from Lord Carter's *Review of Legal Aid Procurement* (House of Commons, 2006) have been unpopular with the legal profession, having been described as 'rigid' and 'complex'. It has also been argued that the reforms have alienated some sections of society most in need of legal services. For example, a single parent working full-time on the minimum wage, and supporting a child, is not eligible for Legal Aid in criminal proceedings at the magistrates' court.

But where do these wider issues about legal aid, legal representation, and access to justice, sit within our examination of fairness in ASBO applications? Is it the case that the ASBO process is largely circumventing safeguards for the protection of defendants and resulting in discriminatory practices targeted, for the most part, at those (traditionally conceived of) 'undesirable persons'? First, let us consider the way in which Galligan reconciles legal representation and fair treatment:

> In determining when legal representation is needed for fair treatment, the principal guideline should be that legal advice and representation are needed when, without them, the person affected would not be able properly to prepare and present his case. ... *The principle of English law is that legal or other representation of a party in an administrative process is not a necessary requirement of procedural fairness. An authority must follow fair procedures, and whether representation is required as a fair procedure depends on the context.* (1996: 365, emphasis added)

Indeed, Adler (2003: 331) talks of 'trade offs' that are made between institutional actors in administrative processes, and, moreover, he recognises a plurality of competing normative positions on what it means to treat people fairly (Adler, 2006: 637). While greater access to the courts in the form of Legal Aid would, in my view, *be a step towards increased access to justice*, it is also essential that the system aims at a balance between accuracy and its cost.

Operating costs are borne by the public in the form of subsidies to the judicial system and by the parties in the form of court fees, solicitor's fess, and litigation costs. However, cost is not the most relevant factor when considering the value of legal representation within the administrative process. Instead, the effects of increased legal provision, particularly with regard to appeal cases, on an overburdened and lethargic summary justice system should be considered. Hence, the effect of increased legal provision on the summary justice system is certainly an important consideration in this context. While there is evidence to show a correlation between legal representation and delay, this is of course no reason to bypass procedural fairness for administrative expediency – indeed, delay may be a necessary factor for good decision-making (Genn and Genn, 1990). However, de facto practical considerations mean that procedural justice is only achieved through an exercise in counterbalance and proportionality. While the enfranchisement of all participants in court processes via, amongst other elements, access to legal representation of a certain standard, should certainly be an aspirational ideal of our system of law, current circumstances necessitate that such objectives are viewed within the wider sphere of an overburdened and lethargic summary justice system.

As a result, jurists and legal professionals/practitioners have argued that fewer cases (relating to lower level criminality and anti-social behaviour such as littering and neighbourhood nuisance) should be addressed by the courts (McInnes, 2005). It is thus highly unlikely that the summary justice system, in its current form, would be able to cope with a significant increase in its case load and it should also be remembered that wide and unfettered access to legal representation would likely give rise to an increase in the number of spurious and/or illegitimate cases.

Nonetheless, current provisions for legal aid are unduly prescriptive and without doubt mean that access to the courts, and consequently to fair treatment, is fundamentally circumscribed for participants in ASBO cases – and civil trials more generally. However, one way of approaching this difficulty without 'opening the floodgates' to unrestricted legal representation is to attempt to ensure fairness from *within* the court process itself. Certainly, given the autonomy entrusted to local enforcement agencies, *a rigorous approach to evidence gathering and case management should be ascribed the highest priority.* Yet, as we have seen from the Study findings, nearly half of solicitors surveyed in England and Wales cited the 'urgent' need for the introduction of case management powers for ASBO applications. A lack of inter-agency consultation and co-operation; inconsistent attitudes towards information sharing; the presence of inexperienced evidence gatherers; the defence rarely serving evidence before trial; vague hearing dates; and, a disjointed framework for the ASBO process with different procedures in different courts, were all contributing factors that solicitors argued necessitated the creation of powers to enable the courts to apply rigorous case management to ASBO proceedings. An illustration of the importance of the requirement for rigorous case management is made by a recent ASBO case in Manchester. In July 2007, a woman in Manchester was awarded £2000 in compensation by her local authority after The Local Government Ombudsman said that Manchester City Council was guilty of an 'abuse of power of nightmarish proportions' in obtaining an ASBO against her based upon false and uncorroborated allegations from a neighbour. In this instance, the court granted an interim order against the woman a week after ASBO papers were served against her.

This case appears similar in kind to examples cited by Halliday (2004) in his study of homelessness decision-making – where legal values were sometimes regarded as 'unwelcome intruders' by authority staff. He illustrates this with reference to the importance of 'professional intuition'. In homelessness decision-making, for example, experienced decision makers develop 'confidence in their ability to gain an almost immediate sense of the truth underlying an applicant's claim for housing so that they are able to "just know" what a case was about'. (p. 54). Halliday identified a 'strong internal culture which resists interference from legal values' (p. 59) so that, in trusting their intuition, decision makers react out of a siege mentality to reject the normative authority of the law (p. 60).

Given the discretion afforded to applicant agencies in the ASBO process, the potential benefits of statutory case management rules for ASBO proceedings undoubtedly merit further consideration and research. Indeed, we have observed that both applicant agencies and the judiciary have *significant* levels of discretion in the ASBO process. Discretion can appear increasingly wide and unfettered, leading to public dissatisfaction with outcomes, when court procedures are themselves inept or lacking in some respect. As Pound (2002 [1931]) has articulated, public dissatisfaction with the courts can indeed be reduced, by alterations in court procedures. In this respect, it is suggested that improved case management and/or the creation of statutory case management powers is a possible means to improve fairness in ASBO procedure(s) and outcome(s). A more rigorous approach to evidence-gathering and case management would go some way to addressing the variation in the standard of evidence required to obtain an interim/ASBO, and the differences in the quality of legal representation available to defendants in ASBO applications which can, it is fair to state, affect the risk of erroneous deprivation of substantive rights in ASBO cases, in respect of defence and appeal procedures.

Interim orders ex parte

The majority of interim orders in Britain are granted on the basis of hearsay evidence or, as the legislation permits, on a *prima facie* basis without the hearing of any evidence or the lodging of any witness statements or productions. Moreover, in England and Wales, interim orders can be granted *ex parte* (without notice to the defendant). It follows, of course, that if an application is made on notice, then the defendant may choose to give evidence at the hearing. However, the Study findings showed that evidence is rarely served at the interim application stage (in either Scotland, or in England and Wales), and as one Sheriff concluded: defence 'lawyers take the attitude that there is so much that has happened before we get to the stage of an anti-social behaviour order that there is no point opposing the interim order'. Consequently, orders that are made *ex parte* necessarily mean that the defendant will have no opportunity to give evidence at this stage of the application process. In terms of the wider implications for procedural justice, we must now consider *ex parte* applications, and applications made on a *prima facie* basis in the context of the right to be heard – and whether such a right does or should exist in relation to ASBO action.

A hearing, in its simplest legal context, is a procedure through which evidence is imparted from both parties, and the process provides an opportunity for argument to be presented from more than one source. In this way, a hearing is important in providing fairness in procedure, and in achieving a balance between the competing interests of the parties. Moreover, good decision-making will most often necessarily require an investigation of an individual's circumstances and a hearing will be an effective means of achieving this. Thus, the opportunity for an individual to be heard within a given legal process is fundamentally bound up with conceptions of fair treatment, impartiality and equity. However, while the principle of a hearing is certainly intrinsic to discussions about fairness and procedure – it is important to consider the boundaries and limits of such a principle, and how it is variously construed. So if we consider the hearing principle in ASBO applications within the wider context of procedural justice, we must first necessarily examine what values are at stake, and what standards they generate in terms of fair treatment of person(s) involved in the legal process.

The hearing process is fundamental to good decision-making and good outcomes, primarily because a hearing embodies the *telos* of the civil procedural system as a 'search for truth'. A hearing also allows the opportunity for an individual to actively advance or to defend their interests, which is a relationship that 'draws on the value of each person being actively engaged in his relationship with the state, rather than being the passive recipient of benefits or the victim of burdens' and consequently, the hearing is 'directly linked to fair treatment' (Galligan,

1996: 349). While the notion of an individual's right to autonomy, self-determination and the preservation of rights in the context of legal process(es) is embodied within the hearing principle, Galligan simultaneously advances another, far less individualised explanation for the right to be heard. Acknowledging that, although one could argue in favour of the hearing principle on the basis of respect – after all, 'respect for a person requires that he be heard' (ibid.) – he is mindful of the fact that 'there is, however, scarce support for this approach in judicial statements of principle' (ibid.). Instead, he posits that:

> The hearing principle might be approached in another way, not in order to establish rights, but to show its value to society as a whole. ... There may also be social gains, perhaps less tangible, in having a citizenry which is active in protecting its own interests and in each being treated with respect by the whole. (p. 352)

Indeed, a consideration of English case law on the hearing principle suggests that the right to be heard is a fundamental tenet of English law. For example, Lord Diplock considered the right to be heard one of the fundamental rights generated by the general duty on administrative officials to act fairly towards those affected by its decisions (*O'Reilly v Mackman [1983] 1 AC 237* at 279). The right to be heard meant in that case learning what is alleged against the person, and then having the chance to put forward an answer to it. Similarly, in Scots law, the Scotland Act 1998, and Article 6 of the Human Rights Act 1998 safeguards the right to 'a fair and public hearing within a reasonable time by an independent and impartial tribunal established by law'. Implicit in the requirement for a 'fair hearing' is the principle of equality of arms between litigants, and the opportunity to present a case. However, while Galligan observes the hearing principle as linked to societal good, and citizen participation, he also recognises that 'the supposed principle that a person should be heard is much less secure as a general legal practice then judicial statements suggest' (p. 355).

In the context of ASBOs granted *ex parte* in England and Wales, it is thus important to note the provision for defendants to make an early challenge to a decision to grant an application made in their absence is included within the relevant statutory provisions. The Justices' Clerks' Society *Good Practice Guide to Antisocial Behaviour Orders* states that: '*ex-parte* interim orders should be given as early a return date as practicable to allow the defendant an opportunity to be heard' (2006: 6). Galligan goes as far as to say that, although '*ex parte* applications by one party in the absence of the other are presumptively unfair...the unfairness can be *removed* by effective procedures for early challenge by the absent party' (1996: 391, emphasis added). While I would not go as far as Galligan in stating that an early return date necessarily *removes* inequity in procedure, the opportunity for a speedy challenge to an incorrect decision certainly provides redress for affected parties. Whether an individual who has had an *ex parte* order issued to them in error[3] would necessarily view an early return date as negating any injustice in the original procedure remains to be seen.

The principle of a hearing – the right to be heard – is thus not by any means absolute in law, and defendants should not necessarily expect to be afforded a hearing. Indeed, the discretionary power of the courts means that they will seek to establish if, given the type of case before them, fair procedure could only be achieved were the defendant given the opportunity to be heard. Galligan states that 'the courts often ask whether a hearing is necessary in the circumstances of the case to ensure fair treatment' (1996: 353). Thus, it is argued that interim orders granted *ex parte* do not *presumptively* infer unfair treatment, rather, the issuing of *ex parte* orders should be considered within the context of the balancing of parties' competing interests. While Galligan contends that 'there is still a strong case for a presumption in administrative processes generally in favour of a hearing, not as a fundamental principle, but for a mixture of practical and value-based reasons' (ibid: 355), it is also apparent that the principle of a hearing falls within the scope of the discretionary autonomy of the judiciary to decide the bounds of 'fair treatment' through their interpretation of symmetry, proportionality and individual rights-based considerations in ASBO applications. Hence the judiciary exists as a 'primary definer' on the use of ASBOs (Donoghue, 2007). Within the ASBO process, law retains its status as a site of power – which is used not simply as a mechanism to circumvent the safeguards of due process, but also to safeguard the interests of ASBO defendants. As one sentencer in the study noted,

> ...the law [on ASBOs] as it currently stands means, I think, that [sentencers] ought to think very carefully about...due process and should give proper regard to that...it's important that justice is not bypassed.

The central question should be therefore – is the balance of interests between public protection and the procedural rights of the defendant correctly struck? It is of course, as we have seen already, important to be careful in distinguishing what is meant by the 'balancing' of rights in ASBO cases since the notion of 'rebalancing' could be interpreted as an explicit reflection of sectional interests (Zedner, 2005). The judiciary remains, despite the existence of a significant degree of discretion, constrained by the existing statutory provisions on ASBOs. While there is the opportunity for law to act as a site of power within which individual autonomy is protected from the coercive influence of disciplinary power, there is also the potential for law to act as an instrument of exclusion and/or domination. Thus, there is *a genuine substantive requirement for improved administrative courts procedures to safeguard the interests of ASBO defendants*. It is certainly the case that there is evidence to illustrate the requirement for revised administrative court procedures which, I have suggested, could be addressed through improved case management. Enhanced case management would contribute to improvements in evidence-gathering and more efficient information sharing, and greater consistency in ASBO administration.

8
Reflections on ASBOs as Social Aegis

It is only a short time ago that commentators had begun to observe what they saw as 'the end of the Respect Agenda' and the relegation of anti-social behaviour as a political priority. The Respect Task Force had been disbanded, and responsibility for anti-social behaviour policy had been split between the Home Office and the new Department for Children, Schools and Families (DfCSF). The numbers of ASBOs issued had significantly declined between 2005 and 2006, and it appeared that anti-social behaviour policy was essentially being politically sidelined. However, as we have seen, the government has recently announced a 'new drive' on anti-social behaviour, and a package of additional measures designed to make the ASBO process 'more efficient'. Although research has been limited, there is also evidence that when ASBOs (and other anti-social behaviour interventions) are being used, they are effective at deterring further anti-social conduct (Home Office, 2008; NAO, 2006).

An attempt to address the problems associated with anti-social behaviour is to be welcomed, since the study of anti-social behaviour touches on important themes of community safety and social cohesion, disorder control, discourse and identity, and social exclusion. However, academic literature on anti-social behaviour management and the use of ASBOs frequently locates the introduction of anti-social behaviour policy within the discourse of a new 'culture of control' (Garland, 2001) which identifies the collapse of penal welfarism and the development of popular punitivist responses to crime and disorder. Indeed, in Chapter 4, we observed the evolution of a new politics of law and order, whereby crime and its control have become increasingly politicised. Collins and Cattermole (2006: 34) suggest that, within this paradigm of 'new penology', the State is no longer the 'sole provider of security' and its attention has become focused upon what Garland (2001) describes as 'adaptive responses'. These responses embody a greater emphasis upon victims, withdrawal of police from the community, and increased community involvement in justice processes; but also *coercive* sovereign state strategies which are concerned with the control and segregation of offenders. As Garland (2001: 17) contends:

> The new infrastructure is strongly oriented towards a set of objectives and priorities – prevention, security, harm-reduction, loss-reduction, fear-reduction – that are quite different from the traditional goals of prosecution, punishment and 'criminal justice'. So while the most prominent measures of crime control are increasingly oriented towards punitive segregation and expressive justice, there is, at the same time, a new commitment, especially at the local level, to a quite different strategy that one might call *preventative partnerships*. Today's most visible crime control strategies may work by expulsion and exclusion, but they are accompanied by patient, ongoing, low-key efforts to build up the internal controls of neighbourhoods and to encourage communities to police themselves.

ASBOs have been identified as part of this new approach to crime control and criminal justice which, as we have seen throughout this book, critics contend embodies marginalisation, exclusion and punitive social control aimed at greater regulation of minority social groups – in particular, the economically disadvantaged. This is commonly the explanation of anti-social behaviour policy and the use of ASBOs that is communicated to students.

Moreover, there appears to be a suggestion within the academic literature on ASBOs, that endorsement or support for anti-social behaviour policy belongs, for the most part, within policy/political or practice spheres. It is not generally acknowledged that there may also be academic 'support' for the ASBO and for its administration/outcomes. It is curious that support or endorsement for anti-social behaviour policy is seen as belonging exclusively within either the political/policy or practice spheres. Perhaps this is because academics are largely concerned with the critical analysis of policy rather than de facto support for it that it is presumed that there will not be any academic voices willingly to display the inferred compliant acquiescence of those 'on-message' local authorities identified as members of the 'enthusiast camp' (Squires, 2008).

One might also detect, perhaps, a propensity within the academic literature for authors to view the academic critique of ASBOs as somehow more worthy, better thought-out and more imbued with moral principle than alternative 'enthusiast' perspectives. Moreover, Nixon and Parr (2006: 79) contend that the 'way in which the problem [of anti-social behaviour] is portrayed in the popular press is...characterised by the use of assertion and a demonising polemic'. This may be so, but it has hard not to see the consanguinity between the 'academic rhetoric' and the 'media rhetoric' on anti-social behaviour – each with their indubitable dismissal of the other's conceptualisation of what anti-social behaviour means, and to whom. In one of the very few critiques of the dominant academic perspective on anti-social behaviour Flint (2006: 331) observes that:

> There is...a strange irony among academic commentaries that criticise anti-social behaviour discourse and policy because of its failure to recognise the diversity of voices and the different ethical interpretations of norms and values while simultaneously dismissing alternative theoretical perspectives...if there is a new politics of conduct (Rose, 1999), then competing notions of what causes and constitutes anti-social behaviour, negotiations about the legitimate rights and responsibilities of individuals, different groups in society, communities, voluntary, private and state sectors and differing perspectives and priorities about how the problem should be tackled are inherently valid and should not be readily dismissed.

In this way, this book has argued that much of the available literature on ASBOs is *ideologically based* – in that the subject of analysis has been, to a large extent, the *culture* of crime and disorder control and not necessarily an empirical analysis of institutions and/or practice. Hence this text has attempted to develop an account of anti-social behaviour policy and the use of ASBOs that provides more *balance* within the academic debate. While I am, as I have stated from the outset, broadly supportive of the ASBO model, and of the principles behind it, it has not been my intention to 'convert' this book's readership into ASBO proponents. Instead, I have sought to present and to evaluate the primary criticisms of the ASBO model – both in terms of arguments made about the theory underpinning the creation of ASBOs, and in terms of criticisms about how ASBOs are working in practice. Let us now return briefly to a consideration of those central criticisms.

Political pragmatism?

A common epistemological assumption of New Labour's approach to crime and disorder is that the introduction of anti-social behaviour policy was politically pragmatic, serving as an electoral sweetener that appealed to voters' law and order sensibilities. Hence, what is worthy of consideration in this context is to what extent New Labour's policy on anti-social behaviour was part of an opportunistic and reactionary political agenda embodying a popular punitive approach to law and order. Was the creation of the anti-social behaviour policy chiefly intended to ensure that – as some commentators suggest – New Labour could be strongly identified as occupying 'law and order' territory, with a 'tough' stance on disorder? This type of argument would seem to suggest that the policy was, at least in part, a product of opportunism and political points scoring. Indeed, it is certainly the case that politicians are vote gluttons and it would be naïve to claim that government policies are necessarily the product of political altruism.

However, as we have seen, the introduction of the policy on anti-social behaviour masked embedded and entrenched obstacles to public protection from anti-social behaviour that had been in existence for some time; whereby the impact and consequences of anti-social behaviour were essentially minimised by justice and welfare systems which was in part manifested by a police reluctance to deal formally with complaints. As Brown's research on anti-social behaviour found, people had previously escaped conviction for anti-social acts 'for two reasons... witness intimidation and... the possibility that the police do not treat anti-social behaviour as "real" crime' (2004: 208). Through their inefficacy, criminal and civil justice systems had essentially devalued the importance of the effects of anti-social behaviour on victims which had subsequently resulted in the germination of the perception that anti-social behaviour did not amount to a legitimate social problem of a level of seriousness requiring positive intervention by the state and its agencies.

Moreover, one of the main arguments cited by local authorities for the introduction of the ASBO was that it was able to address the *cumulative effect* of anti-social behaviour. Evidence had shown that individual offences were often minor when viewed on their own, carrying limited sentences, and prosecution achieved little because the root cause was the cumulative effect of the behaviour and not the impact of individual incidents. Case law had also demonstrated there to be intrinsic difficulties for those seeking remedies as tenants, and while civil injunctions could be pursued through the courts; the process had proved to be both expensive and resource intensive (Ashworth, 2004). By this token, there was an inherent dissatisfaction with the criminal law as a means of addressing persistent acts of anti-social behaviour. The criminal justice system had demonstrated itself to be conspicuously ineffective in dealing with certain forms of criminal and sub-criminal behaviour particularly in areas of concentrated multiple deprivation – where poor schools, inadequate health services, vandalism, graffiti, social tension, poverty, drug use and dealing, and high dependency ratios were commonplace (Gordon and Pantazis, 1997; Hills et al., 2002).

Indeed, empirical evidence exhibits the manifestly erroneous social conditions displayed in both inner city and suburban areas of deprivation. Residents cited not only problems related to serious crime but also the pernicious, negative cumulative effect of persistent petty offending and disorder (Home Office, 1992). Research also associated anti-social behaviour with crime and the fear of crime. While anti-social behaviour can be located as a problem affecting all housing tenures (Scott and Parkey, 1998) its impact is most pervasive in inner city areas with high density social housing. Hence, the criminal justice system, acting as the facilitator for this polemic, was demonstrably *unsatisfactory* in addressing persistent and entrenched criminal and sub-criminal behaviours. Consequently, the ASBO was designed to act as a criminogenic millstream – at once unencumbering the criminal justice process, freeing it from legislative and systemic restraints and barriers to efficacy, and enabling it to regain functionary autonomy. In effect, ASBOs were to be a new genus of preventative order with 'public protection [as their] goal' (Ashworth, 2004: 267). Thus, ASBOs were not created simply as a product of political opportunism to enable New Labour to monopolise 'law and order' territory, they were a political response to an identified problem which existing socio-legal structures and interventions had failed to adequately address.

Community (in) action

However, a number of critical commentators have argued that the creation of the ASBO as a (punitive) statutory intervention to deal with community disorder represents a misguided and erroneous attempt to address anti-social conduct of the type which is in fact best addressed through *informal* modes of regulation. The perception that anti-social behaviour should be a matter dealt with informally within communities and between neighbours is, as we have seen throughout, a perception which continues to prove considerably resistant to change. For example, echoing the suggestions of Burney (2005), Simester and von Hirsch (2006: 176) also advocate greater community involvement in issues of neighbourhood disorder *as opposed to* increased statutory controls. They suggest that 'it may also be worthwhile to...direct greater resources toward fostering a more vocal community, one better placed to resist intimidation. To the extent that these solutions are effective, they seem preferable to the much more radical and problematic device of the [ASBO]'. Again, it is important to consider how realistic, practicable, or indeed, safe, it is for individuals and communities to bear responsibility for addressing acts of anti-social behaviour within their own neighbourhoods. Of course, a community that is strongly cohesive is undoubtedly better equipped to reinforce acceptable norms of behaviour but it is more difficult to predicate the ways in which communities can *effectively* challenge violence, harassment and intimidation. Many communities that suffer from acts of anti-social behaviour are already strongly cohesive – and are communicative about what levels of behaviour they are willing to 'tolerate' – but this does not mean that they are necessarily better placed to deal with acts of intimidation and aggression. Tellingly, Simester and von Hirsch (2006: 193) clearly see ASBOs as not sustainable 'as proper forms of criminal prohibition'. In order for ASBOs to become legitimate they would, they suggest, have to become the equivalent of civil ancillary orders which would consequently *significantly restrict their scope*. However, practitioners and policy-makers have both argued that any reduction in the scope of the ASBO would necessarily impact substantially upon its effectiveness in addressing anti-social behaviour(s) (House of Commons, 2005).

Social control

As we have seen in the course of this text, while the purpose of anti-social behaviour policy was to facilitate increased access to (both civil and criminal) legal remedy for victims of anti-social behaviour, critical commentators have identified the ways in which it has also simultaneously facilitated a plethora of greater forms of (social) control. Control is of course indubitable in anti-social behaviour policy; however, policy does not necessarily involve *negative* forms of (centralised) social control at every level. Implementation is pivotal: as we have seen, the opportunity for discretion within policy and legislation means that there will be disparity and variational spread in the strategies that local authorities employ to address anti-social behaviour proportionately (Donoghue, 2005; 2007). Moreover, the discretion afforded to the courts is instrumental in promoting the collective, social values *of law* (in respect of the protection of vulnerable individuals/communities who are victims of anti-social behaviour) but also in protecting the *individual* from the over-extending autonomy of the State.

Anti-social behaviour policy has the potential to be at once a mechanism for control in the pejorative sense but also, contingent upon implementation, it can form the basis of a type of 'beneficial' control which is neither the product of state coercion nor a form of control that is simply being passively tolerated by affected individuals. Rather, the control may be used to limit or to curb harmful behaviours, the effects of which are to be found most prevalently in areas of concentrated multiple deprivation (Millie et al., 2005). In this respect, the implementation of policy can induce consequences that are *desired* or even *encouraged* by particular affected groups who are engaged reflexively in contributing to and shaping policy and practice outcomes (Lianos, 2003). As such, anti-social behaviour risk assessment and interventions of 'control' may contribute to an individual's sense of (ontological) security in ensuring that an 'individual feels free from anxiety about the existence, extent and stable reproduction of the objective or intersubjective conditions of his or her security' (Loader and Walker, 2007: 158).

Similarly, Arthurson and Jacobs (2006: 276) have observed 'a tendency for writers drawing from Foucauldian perspectives to undervalue the demand for action by local residents or other individual actors'. Harrison (2001: 66–7) also recognises that: 'social order practices are not simply a product of the power or needs of employers, ruling elites or capital, but reflect social divisions within class groupings, as well as the interests and actions of those who feel the need for defence or confirmation of their lifestyles. Thus grass roots agency should not be overlooked'. And it is the experiences and the engagement of local residents that should be the primary focus of interest here. It was observed by Chadwick L.J. in *Northampton BC v Lovat [1997] 96 LGR, 548* that

> ...those who live or work on a council estate and are affected by the conduct of council tenants on that estate will expect the council to do something about it. The housing department will receive complaints which will have to be addressed....

To present a picture of anti-social behaviour interventions (in social housing areas particularly) as wholly a matter pertaining to the unsolicited *imposition* of punitive measures of social control upon an entire social group is in part to ignore and/or to disregard both the extent to which these forms of control are supported by individuals and/or communities, but also the extent to which groups/individuals actively engage in these new social processes. Indeed, in Chapter 5, it was observed that the emphasis placed upon community governance and engagement with anti-social behaviour policy interventions has resulted in the *empowerment* of communities (Flint, 2006). In this way, 'empowering residents individually through facilitating and supporting individuals to report incidents, and collectively through community mediation techniques and community consultation... are important mechanisms here' (Casey and Flint, 2008: 114). As such, there is evidence that ASBOs are capable of enabling a *positive* process of engagement among local authorities, the police, housing professionals and residents.

Due process

One of, if not the most important criticisms that has been made of the ASBO model has been the way in which procedural safeguards traditionally associated with the criminal trial have been circumvented by virtue of the hybrid civil/criminal status of the order (Ashworth, 2004; Ashworth and Zedner, 2008). In fact, even before the creation of the ASBO, the difficulties associated with maintaining due process safeguards when using interventions to tackle disorder were observed. Although their 'broken windows' thesis on disorder essentially embodies a zero-tolerance approach to incivilities, Wilson and Kelling (1982: 36) nonetheless recognised the difficulties inherent in such a subjective process of disorder control:

> None of this is easily reconciled with any conception of due process or fair treatment.... We have difficulty thinking about such matters, not simply because the ethical and legal issues are so complex but because we have become accustomed to thinking of the law in essentially individualistic terms. The law defines my rights, punishes his behaviour, and is applied by that officer because of this harm. We assume, in thinking this way, that what is good for the individual will be good for the community, and what doesn't matter when it happens to one person won't matter if it happens to many. Ordinarily, those are plausible assumptions. But in cases where behaviour that is tolerable to one person is intolerable to many others, the reactions of the others – fear, withdrawal, flight – may ultimately make matters worse for everyone, including the individual who first professed his indifference.

Indeed, in the course of this text, notions of community/individual tolerance and intolerance have been discussed, and while the argument progressed here has been that although 'anti-social behaviour' is a subjective term, it is not so relative as to preclude a duty to behave reasonably to one's neighbour(s), it has also been argued that because the use of ASBOs is contingent upon a subjective criteria – it necessitates rigour and standardisation in legal and court processes.

In Chapters 6 and 7, the problems associated with the ASBO court process were highlighted. In terms of *victims* of anti-social behaviour, it was found that undue delay in the court process, victim intimidation, and a lack of support for victims and witnesses were all problematic issues. As part of the government's 'new drive' on anti-social behaviour, the Home Secretary has prioritised setting maximum waiting times and limiting the number of times cases can be adjourned. In those cases where witness intimidation is a serious problem, local authority staff are to be trained in the ASBO court process so that they can present cases where necessary. There is also to be more counselling and support available for victims of anti-social behaviour. These are positive developments for victims of anti-social behaviour – since evidence shows that many victims suffer persistent harassment, intimidation and damage to their property while an applicant authority is attempting to obtain an ASBO (Home Office, 2009).

However, as we have observed in the course of this book, there are also legitimate concerns about ASBO court procedure and its (insufficient) attention to due process. In previous chapters, processes of evidence-gathering, interim orders granted *ex parte*, the defence of ASBO action, and the prosecution of ASBO breaches have all been identified as problematic. While the additional measures proposed by the Home Secretary continue to be concerned with protecting *victims* of anti-social behaviour – and making the ASBO process quicker and more efficient – there has been a conspicuous lack of interest in how the ASBO process might be adapted to address concerns about evidential requirements and procedural protections arising out of the hybrid civil/criminal status of the ASBO.

Focusing upon the rights of the defendant in ASBO applications does not by implication ignore the rights of the victim in ASBO cases. Indeed, if the positive function of law as a 'vital regulatory mechanism' as well as 'a source of individual empowerment' (Sommerlad, 2004: 350) is to be discharged, it is necessary that identified problems in the current system are understood and addressed. If ASBOs are to be continued to be used as a preventative (and protective) order in the future, and not repealed or replaced by future governments, then their use must be legitimate, and it must *be seen to be* legitimate. A procedure that creates civil orders that are illegitimate and ineffective will no doubt be replaced or removed – and so ASBOs must evolve, they must become fairer in their application, their quantifiable effectiveness must be demonstrated by further empirical research, and *the conflict between protecting individuals from anti-social behaviour versus the rights of defendants* must be more adequately addressed. As such, I have advocated the need for improved pro forma legal procedure(s) in order to negate inequality in the formal administration of law.

Civic reciprocity

In previous chapters, we have considered: how far do we inculcate the notion of responsibility and personal accountability into our analyses of anti-social behaviour? Are economic and social disadvantage mitigating factors in accounting for anti-social behaviour? Or should personal choice feature as a broader philosophical underpinning for such accounts of anti-social conduct? The criminologist Richard Wright maintained that free will and rational choice orientations are 'often viewed by other criminologists as being conservative. If you argue or advocate a free will or rational choice approach, many will say "Well, you don't take into consideration larger structural issues," and it's immediately assumed that you are a conservative [criminologist]' (quoted in Miller, 1996: 111). And indeed, this argument goes to the heart of many (academic) conceptualisations of anti-social behaviour policy and the use of ASBOs. In Chapter 3, we observed the way in which contemporary law and order policy is often identified in the criminological literature as 'Right leaning' or 'Centre-Right' in its ideological underpinnings. By this token, anti-social behaviour policy is characterised as predominantly concerned with enforcement and as such is seen as disinterested in addressing social exclusion and disadvantage. Hence this type of academic critique understands ASBOs as part of broader socio-political agenda progressed by majority social groups and ruling elites, aimed at 'more effectively regulating the poor'. In Chapter 5, we then observed a pervasive reluctance (within the academic debate) to determine a causal link between behavioural conduct and choice. As such, this facet of critique argues that current policy on anti-social behaviour *reduces* behaviour to a moral choice and ignores the importance of socio-structural contexts (McIntosh, 2008).

For the purposes of my arguments in this book, this aspect of the critique on anti-social behaviour is the one that I find most problematic. Requiring personal responsibility for individual conduct (and its effect on others) does not infer a de facto de-contextualisation of a perpetrator's social circumstances but it does infer that there are standards of behaviour that it is legitimate to expect of everyone, regardless of their socio-economic status. Undoubtedly, the ambiguity of the statutory definition of anti-social behaviour means that procedures for obtaining ASBOs should be ameliorated to ensure that the orders are not granted inappropriately and inconsistently. Moreover, Mayfield and Mills (2008) suggest that there is a need to agree standard definitions of anti-social behaviour across the country if we are to record and to measure anti-social conduct in the same way. However, for the most part, the conduct that forms the basis of anti-social behaviour orders such as vandalism, graffiti, aggressive behaviour, excessive noise pollution, drug dealing and so on, it is fair to say, might legitimately be regarded as 'unreasonable'. In this way, while what is considered anti-social conduct necessarily fluctuates between locales/individuals, there still remains a standard of behaviour which might properly be regarded as unreasonable and thus requiring (preventative/punitive) intervention. Rejecting a link between behaviour and choice not only goes some way to abdicating responsibility for unreasonable behaviour, but seems to infer that behaving anti-socially to one's neighbour's can be rationalised and then justified by virtue of mitigating socio-economic circumstances. Personal biographies and individual contexts are of course important in explaining/understanding behavioural conduct but should not be used as factors to devalue the existence of a substantive link between conduct and choice, and above all, the primacy of personal responsibility.

While it is certainly the case that early intervention and strategies of support to rehabilitate ASBO recipients should be integral to the policy framework, and moreover, that the causes of social and economic inequality should be addressed, these are not in themselves precursors to reasonable behaviour. It should be a priority to address social exclusion and disadvantage in all its manifestations, but equally, it should be a priority to deal with (immediately occurring) anti-social behaviour and the very detrimental effects that it can have on the lives of individuals and groups of residents. The two are not mutually exclusive. To this effect, the Home Secretary, Alan Johnson, has argued:

> In today's world, it sometimes seems that Home Secretaries are presented with a stark choice – between siding with those affected by crime and anti-social behaviour, and tackling the underlying causes of criminal activity. It's presented as if somehow, any drift towards the latter undermines the former. But I don't believe this is the

case. There is no inherent contradiction between the need to bring offenders to account and protect people from the immediate dangers of crime, and the need for longer term strategies to prevent criminal activity. (Home Office, 2009)

Of course, it would be better if we did not live in a society which *enforced* civic reciprocity and reasonable behaviour, but it is the case that anti-social behaviour perpetrated by a minority (for the most part in areas of economic disadvantage) has a very detrimental impact upon the lives of others. While the argument that the focus of anti-social behaviour policy has broadened to incorporate a more general preoccupation with nuisance and the intolerance of incivility, undoubtedly necessitates improved standardisation in ASBO administration to guard against their use as tools of prejudice and discrimination (Donoghue, 2007), it remains the case that evidence demonstrates that anti-social behaviour has a serious impact upon the lives of a minority of the population and it is the argument of this text that the ASBO represents a serious attempt to address this problem.

Critics suggest that defining what is anti-social conduct is a relative project, generally premised upon 'middle-class' standards of normative behaviour. Moreover, as we have seen, others posit that attempting to define or to quantify 'shared values' or a 'common good' is in it self a redundant exercise given the cultural relativity of these terms (Collins and Cattermole, 2006). Indeed, the notion of civic reciprocity (and its viability as a concept) is fundamental to the debate about anti-social behaviour policy. Politicians and policy-makers have advocated the need for a culture of 'Respect' and for the pairing of rights and responsibilities, while opponents regard these notions as naively construed and normatively suspect. This is an inherently ideological debate (although proponents of each argument would cite empirical evidence of the in/existence of 'civic or shared values' in support of their view). It is because discussion of anti-social behaviour is so ideologically embedded that analyses of anti-social behaviour policy must seek to represent both theory and practice. It has been the contention of this text that too much existing scholarship on anti-social behaviour is constrained by normative ideological biases and too little representation of alternative theoretical and practice perspectives.

That ASBOs are part of a sinister, punitive agenda of control which deliberately seeks to marginalise and to exclude 'targeted populations' is an argument which this book flatly rejects. However, there are without doubt other legitimate concerns (discussed in the previous chapters) to be had about ASBO legal and court processes and the way in which due process safeguards may be being circumvented. These are important concerns, but they do not necessitate the abandonment of the ASBO model – merely its revision and improvement. Enhanced case management to ensure improvements in evidence-gathering and information exchange between agencies, and the standardisation of ASBO administration to ensure greater consistency in applications is warranted. However, this book rejects those arguments which contend that New Labour policy on anti-social behaviour has aggravated what was essentially a small problem resulting in more individuals becoming 'more dissatisfied with their lives' (Tonry, 2004: 57). While anti-social behaviour seriously affects the lives of a 'minority' of the population, this does not in effect make anti-social behaviour a de facto small problem. Moreover, research has found that anti-social behaviour 'has little or no effect on the lives of the majority of the population' (Millie et al., 2005: 1) so claims that anti-social behaviour policy has substantively inflated perceptions about its existence and has made more people dissatisfied with their lives seems somewhat tenuous. The effect of anti-social behaviour upon individuals (particularly those resident in inner cities and in areas of deprivation where anti-social behaviour is most prevalent) is serious and must not be minimised. Hence, it has been argued in this text that the introduction of ASBOs should not be understood as part of a 'culture of control' but as underpinned by a process of social protection, of 'social aegis'. I would suggest that rather than beginning analyses of ASBOs from a perspective that identifies anti-social behaviour management as vicariously linked to oppression and marginalisation, it is worth considering Tony Blair's rationale for the introduction of the policy as a starting point for analyses:

> This is not a debate between those who value liberty and those who don't. Critics need to answer the following question: if the criminal justice system was failing people, as it clearly was, what ought we to have done? To do nothing is one option. But surely it is to do better by the British people to devise relevant powers, limited by the right of appeal, to ensure that communities do not have to live with unacceptable levels of fear and intimidation. No liberal democracy can countenance the tyranny of a minority in any of its communities.

When we think about anti-social behaviour management and the use of ASBOs, there are numerous important and difficult questions to consider, emanating from ideology, political philosophies, policy and practice, implementation, and beyond. What must be wary of, however, are those reductive characterisations of ASBOs which identify anti-social behaviour management as simply a technique of social control, designed by those who wish to achieve better regulation of the poor. Not only are such ad hominem arguments misrepresentative but they subvert the practical reality of anti-social behaviour for its victims. The reality is that anti-social behaviour is a genuine and serious problem in Britain. While it is experienced by a minority of the population, its effects are pernicious and debilitating. We must never lose sight of this when we consider ASBOs in the context of a 'culture of control'.

Notes

2 The End of Respect?

1. Johnson was made the general secretary of the Union of Communication Workers in 1993 and was the only union leader to support Tony Blair in his campaign to abolish Clause IV of the Labour party's constitution, which called for 'the common ownership of the means of production, distribution and exchange'. Clause IV was officially replaced in 1995 and is seen as the seminal moment at which Old Labour became New Labour.

2. Although it is interesting to note that, shortly after his appointment, in what was seen as a significant move, Johnson abolished government plans to make ID cards *compulsory*. He has also hinted at plans to make possible the removal of some minor convictions from the Police National Computer (PNC). As such, commentators have suggested that Johnson has distanced himself from the positions adopted by a succession of other Home Secretaries who had continued to support the introduction of compulsory ID cards and other new 'security' measures that were often criticised for being associated with greater state surveillance and control – the 'Big Brother' culture.

3 Anti-Social Behaviour: The Political Landscape

1. See the 'Respect' website at http://www.respect.gov.uk/members/article.aspx?id=7536

2. In Scotland, anti-social behaviour is defined as 'behaviour which causes or is likely to cause alarm or distress', under s.143 of the 2004 Act.

3. In Scotland, however, ASBO duration is a matter for the discretion and evaluation of the individual Sheriff presiding over the application.

4. Part 1, s.1 (1) of the 2004 Act, requires that 'each local authority...shall... prepare a strategy for dealing with antisocial behaviour.' Section 17 of the 1998 Act places a statutory duty on chief police officers and local authorities in England and Wales to work together to develop and implement a strategy for reducing crime and disorder.

5. Given that proposals for a clearer definition of anti-social behaviour are 'beset by the hopelessness of trying to define an umbrella term like "antisocial behaviour" precisely' (2003: 206), Macdonald has subsequently proposed two new clauses to section 1(1) of the 1998 Act with the objective, not of defining 'anti-social behaviour', but of expressly detailing the *principles* underpinning the ASBO, in order to provide clarity and consistency in application. Specifically, he suggests that section 1(1)(a) be qualified to require that the behaviour complained of was persistent, that it caused a serious level of harassment, alarm or distress, and that the perpetrator *intentionally* committed the anti-social acts.

6. An instructive comparison can be made here with Innes' influential work on 'signal crimes' (2004); see also Pawson and Mckenzie, 2006; Sennett, 1996; 2003.

7. For instance, drug use/dealing is more frequently cited in social housing areas, while litter and graffiti are identified more often in affluent urban areas.

8. This is an issue that has generated problems between, for example, London boroughs and other authorities over accommodating tenants, (Taylor, G. [of the Local Government Association], The Guardian, 'Home and Away', 23 May 2001).

9. The Metropolitan Police Authority, in a 2001 report by the Commissioner, stated that a detective sergeant involved in one application had spent more time on an ASBO application than she would normally have spent on a very serious criminal investigation. Officers estimated the cost of obtaining this ASBO was in excess of £100,000.

10. In 2004, the minimum estimated cost was £150 (for an ASBO on conviction). The maximum estimated cost was £10,250, which was the first ASBO issued by one CDRP. ASBOs on conviction were generally cheaper to administer than other types of ASBOs, costing on average £900 compared with over £3000 for stand-alone orders (Home Office, 2004b).

11. In 2005, only 19 local authorities in Scotland were found to be collecting financial information about the cost of using ASBOs and other measures to tackle anti-social behaviour (Scottish Executive, 2005a: para. 2.4). Previously, the Convention of Scottish Local Authorities (COSLA) had emphasised the difficulties in estimating the average costs of ASBOs, but suggested that the costs were more often between £5000 and £20,000 (Scottish Parliament, 2004).

12. I have conducted interviews with anti-social behaviour unit managers in Scotland as part of my research on ASBOs. The study and its methodology are discussed in Chapter 7.

13. Subsequently, in the details that emerged of the toddler's murder, it was disclosed that the two boys had battered Jamie Bulger to death with bricks and an iron bar, and then left his body on a railway track to be cut in half.

4   Anti-Social Behaviour: The Historical Landscape

1. For example, Millie et al. found that neighbourhood residents frequently regarded anti-social behaviour 'as a symptom of social and moral decline' (2005: 7), while Sennett (1996) and Innes (2004), amongst others, have observed the contribution of media discourse to current perceptions about the nature and incidence of anti-social behaviour.

2. For an interesting discussion of the evolution of 'chav' culture in late-modern society, see Hayward, K. J., and Yar, M., (2006) 'The "Chav" phenomenon: consumption, media and the construction of a new underclass', *Crime, Media, Culture*, 2 (1) 9–28.

5   Anti-Social Behaviour and Social Housing

1. It should be noted, however, that the findings of the 'one day count' study are not unanimously accepted. Burney (2005: 81), for example, has argued that because there was no recognition of the potential for 'double-counting'

by different authorities, and furthermore that measurements were not incorporated into the study which could ascertain the differing degrees of upset/harm caused by the behaviours to individual complainants – particularly in view of the likelihood that some individuals are more likely to complain more frequently than others, who may do so rarely or never – that 'it cannot be held that either quantitatively or qualitatively this was a meaningful exercise.' However, in view of the substantial amount of research data on the economic cost of anti-social behaviour, one could, nonetheless, certainly arrive at the legitimate conclusion that anti-social behaviour is very costly, both in terms of costs to the community, and costs to the individual.

6 ASBOs in Practice

1. For the purposes of the study, 'outcomes' was defined as the result of an ASBO court application. That is to say, 'outcomes' spanned a range of consequences: whether or not an application was successful; to what extent an order was amended (in respect of prohibitions and duration) before being applied; and the court's approach to an application for breach proceedings. 'Outcomes' did not, however, refer to the *effectiveness* or otherwise of an ASBO being served – it referred only to the result of the application process.

2. Part 1 of the Anti-social Behaviour Etc. (Scotland) Act 2004 places a statutory duty on each local authority and relevant chief constable in Scotland to prepare a strategy for dealing with anti-social behaviour in the authority's area. Similarly, the Crime and Disorder Act 1998 also places a statutory duty on chief police officers and local authorities in England and Wales to work together to develop and implement a strategy for reducing crime and disorder, hence individual authorities in both Scotland, and in England and Wales, possess anti-social behaviour co-coordinators, community safety officers, and so on, for the purposes of their statutory duties in respect of reducing anti-social behaviour, crime, and disorder.

3. However, under s.1E of the Crime and Disorder Act 1998, the police and local authorities must consult each other when applying for orders.

4. Unlike in Scotland, where it is only the local authority or Registered Social Landlord (RSL) which acts as the relevant agency for the purposes of ASBO applications, in England and Wales, a relevant authority can be a local authority, registered social landlord (RSL) or the local police force. As a result, many local authorities in England and Wales are not involved as the lead agency in pursuing ASBO applications for their area. Instead, applications are exclusively applied for by the local police force. However, under s. 1E of the Crime and Disorder Act 1998, the police and local authorities must consult each other when applying for orders.

5. Anti-social Behaviour Act 2003; Anti-social Behaviour Etc. (Scotland) Act 2004.

6. Although the House of Lords had previously set out the law on the standard or proof in respect of ASBO applications in *McCann*, the position was not binding in Scotland. Therefore, Scottish courts were not obligated to follow the House of Lords judgement. Subsequently, the standard of proof applied in Scottish cases is, in contrast to cases in England and Wales, the civil standard of proof – and not the heightened civil standard (equivalent to the criminal

standard) that is applied South of the border. This created an amount of uncertainty and confusion among the legal profession (in Scotland) as to the appropriate standard of proof required in ASBO cases. For example, in *Glasgow Housing Association Ltd v O'Donnell (2004)* GWD 29–604, Sheriff Holligan considered the criteria to be satisfied before an interim ASBO could be granted. It was held that the court had to be satisfied that the interim order was 'necessary' to protect relevant persons from further anti-social acts or conduct. However, in *Glasgow Housing Association Ltd v Sharkey (2004)* HousLR 130, Sheriff Principal Bowen commented on the judgement of Sheriff Holligan in *O'Donnell*, and concluded that Sheriff Holligan's observation that the necessity test was a 'high' one went too far. Sheriff Principal Bowen decided that 'necessity' was a matter of fact to be decided on a case by case basis, which was an exercise of judgment for the presiding Sheriff in an individual case. Sheriff Principal Bowen's judgement was confirmed in *Aberdeen City Council v Fergus (2006)* GWD 36–727, whereby Sheriff Principal Young confirmed that in considering whether an interim order should be made, the court undertook a two-stage test. The court requires to be satisfied, first, that the person was engaged in anti-social behaviour, and second, that an interim order should be made. In looking at this matter, no particular standard of proof is applicable. The second stage requires the Sheriff to consider all relevant matters, ignore irrelevant matters, correctly apply the law and come to a decision which is reasonable. However, it is apparent that there exits inconsistency in the courts with regard to the standards that are required by Sheriffs for successful interim/ASBO applications. In Fife, for example, interim orders can be sought and are granted in Chambers without the need for a court hearing. Yet, in Glasgow, it has been reported that interim order applications have been rejected until a full proof submission is made. Moreover, in some areas, the evidential requirements laid down by Sheriffs for interim orders are 'little different from what was deemed necessary to justify applications for full ASBOs' (Scottish Executive, 2005a, s.2.21).

7. This is a special form of injunction stopping a party from disposing of assets or removing them from the jurisdiction (out of the country).

8. An Anton Pillar Order directs a defendant to disclose or to deliver up documents

9. Although England and Wales, and Scotland have separate legal systems, the courts in all three jurisdictions have long expounded the general principal that both legal procedures and the decision-making of officials must be fair – and that a duty exists to act fairly and to afford all participants the right to be heard.

10. For example, interim/stand-alone order application stage, breach of an interim/stand alone order, order on conviction application, breach of an order on conviction, and so on.

11. This legal position is not unique to ASBOs. For example, see restraining orders (under the Protection from Harassment Act 1997, s.3) and sex offender orders (under the Sex Offenders Act 2003 and the Crime and Disorder Act 1998. s.2).

12. An order must be 'necessary' for the protection of persons from behaviour likely to cause harassment, alarm or distress.

13. ASBIs are also civil orders, designed to prevent and to control anti-social behaviour. Obtained in the County Court, an ASBI can compel a person over the age of 18 to do something and/or prevent a particular type of behaviour/action. Breach of an ASBI remains a civil court procedure however, and the court can impose a fine or a period of imprisonment. Using their powers under s222 of the Local Government Act 1972, local authorities can apply to the civil courts for injunctions to restrain anti-social behaviour that constitutes a public nuisance.

14. In her early research on ASBOs, Campbell (2002: 56) had found that the average length of time, from summons to final hearing, was 13 weeks, with some applicant agencies reporting up to 6 months.

15. Campbell (2002: 35) had originally found that, in 71 per cent of cases studied in England and Wales, an external lawyer presented the ASBO case in court, and only 10 per cent of cases were presented by a police force or local authority affiliated lawyer. However, this study's survey responses found that 76 per cent of cases were presented by internal counsel. This is most likely due to the increased level of experience and the greater confidence of internal counsel in preparing and presenting ASBO applications in court.

16. The prohibition of criminal behaviour is an aspect of ASBO use that Sheriffs felt very strongly about. It is suggested that this issue is, however, specifically pertinent to Scotland, and the Scottish Courts, because in England and Wales the police are empowered under the Anti-social Behaviour Act 2004 (as amended) to act as a 'relevant authority' for the purposes of ASBO applications.

17. Although, in Scotland, the granting of an interim ASBO does not allow a local authority/RSL to convert a Scottish Secure Tenancy (SST) to a Short Scottish Secure Tenancy (SSST), such a right only exists in relation to the granting of a full ASBO.

18. The conditions for the making of an order on conviction are the same as those for the making of a full order, with the exception that the words 'relevant persons' in s.1(1)(b) of the Act are replaced with the words 'persons in any place in England and Wales'.

19. Although in *R v W and F [2006] EWCA Crim 686*, the Court of Appeal set out general guidance on court procedure for orders on conviction. It is important to note also, that the granting of an order on conviction is conditional upon the prosecution being able to demonstrate anti-social behaviour by the defendant, in addition to the condition of 'necessity'. However, the impact of the sentence on the 'necessity' for an order should also be considered, since one may make the other unnecessary (JSB, 2007: 36).

7   ASBOs and the Targeting of 'Undesirable' Persons

1. Although not ideal, the opportunity for poor decision-making is somewhat mitigated by an opportunity for swift redress in a full hearing.

2. The lack of conclusive evidence of the 'effectiveness' of ASBOs has been noted. However, the findings of the National Audit Office's study (2006)

which suggest that ASBO interventions can be effective in reducing anti-social behaviour are also acknowledged.

3. Perhaps, as suggested by the European Commissioner on Human Rights, as a result of a spiteful neighbour(hood)/community vendetta against them (Gil-Robles, 2005).

References

Accounts Commission (1997) *A Safer Place: Property Risk Management in Schools.* Edinburgh: Accounts Commission for Scotland.

Adler, M. (2003) 'A socio-legal approach to administrative justice', *Law and Policy*, 25(4): 323–52.

Adler, M. (2006) 'Fairness in context', *Journal of Law and Society*, 33(4): 615–38.

Armitage, R. (2002) *Tackling Anti-social Behaviour: What Really Works.* London: NACRO.

Armstrong, D. (2004) 'A risky business? Research, policy, governmentality and youth offending', *Youth Justice*, 4(2): 100–16.

Arthurson, K. and Jacobs, K. (2006) 'Housing and antisocial behaviour in Australia', in J. Flint (ed.) *Housing, Urban Governance and Antisocial Behaviour.* Bristol: Policy Press.

Ashworth, A. (1998) *The Criminal Process: An Evaluative Study.* Oxford: Oxford University Press.

Ashworth, A. (2004) 'Social control and antisocial behaviour: The subversion of human rights?' *Law Quarterly Review*, 120: 263–91.

Ashworth, A., Genders, E., Mansfield, G., Peay, J. and Player, E. (1984) *Sentencing in the Crown Court: Report of an Exploratory Study.* Oxford: Oxford Centre for Criminological Research.

Ashworth, A. and Zedner, L. (2008) 'Defending the criminal law: Reflections on the changing character of crime, procedure, and sanctions', *Criminal Law and Philosophy*, 2: 21–51.

Atkinson, R. (2006) 'Spaces of discipline and control: The compounded citizenship of social renting' in John Flint (ed.) *Housing, Urban Governance and Anti-social Behaviour.* Bristol: Policy, 99–106.

Bannister, J., Fyfe, N. and Kearns, A. (2006) 'Respectable or respectful? (In)civility and the city', *Urban Studies*, 43(5/6): 919–37.

Beck, U. (1992) *Risk Society: Towards a New Modernity.* London: Sage.

Beck, U. (1999) *World Risk Society.* Cambridge: Polity Press.

Bimrose, J. (2004) 'Sexual harassment in the workplace: An ethical dilemma for career guidance?' *British Journal of Guidance and Counselling*, 32(1): 109–21.

Blair, T. (1998) *The Third Way: New Politics for the New Century.* London: The Fabian Society.

Blau, J. R. and Blau, P. M. (1982) 'The cost of inequality: Metropolitan structures and violent crimes', *American Sociological Review*, 83: 114–29.

Boas, F. (1974) 'The principles of ethnological classification', in G. Stocking (ed.) *In a Franz Boaz Reader: The Shaping of American Anthropology.* Chicago, IL: University of Chicago Press, 61–7.

Boethius, Ulf. (1994) 'Youth, media and moral panics', in Johan Fornas and Goran Bolin (eds) in *Youth Culture in Late Modernity.* London: Sage, 39–57.

Bond-Taylor (2005) 'Political Constructions of the Anti-social Community: Developing a Cultural Criminology', Paper presented at the HSA Autumn Conference 8–9 September 2005.

Bowling, B. (1999) 'The rise and fall of New York murder: Zero tolerance or crack's decline?' *The British Journal of Criminology*, 39: 531–54.

Brand, S. and Price, R. (2000) *The Economic and Social Costs of Crime. Home Office Research Study 217*. London: Home Office.

Braswell, M. C. and Whitehead, J. T. (1999) 'Seeking the truth: An alternative to conservative and liberal thinking in criminology', *Criminal Justice Review*, 24(1): 50–62.

British Institute for Brain Injured Children (BIBIC) (2006) *Research on ASBOs and Young People with Learning Difficulties and Mental Health Problems*, available at http://www.bibic.org.uk/newsite/general/pdfs/ASBOandYOTsummary.pdf [accessed 2 September, 2006].

Brown, A. (2004) 'Antisocial behaviour, crime control and social control', *The Howard Journal of Criminal Justice*, 43(2): 203–11.

Burney, E. (2002) 'Talking tough, acting coy: Whatever happened to the anti-social behaviour order?' *The Howard Journal of Criminal Justice*, 41(5): 469–84.

Burney, E. (2005) *Making People Behave*. London: Willan.

Burney, E. (2008) 'The ASBO and the shift to punishment' in *ASBO Nation: The Criminalisation of Nuisance*. Bristol: Policy Press.

Calder, G. (2003) 'Communitarianism and New Labour', *The Electronic Journal: Social Issues*, November 2(1).

Caldwell, M. (1999) *A Short History of Rudeness: Manners, Morals and Misbehaviour in Modern America*. New York: Picador.

Campbell, S. (2002) *A Review of Antisocial Behaviour Orders (Home Office Research Study No. 236)*. London: Home Office.

Card, P. (2006) 'Governing tenants: From dreadful enclosures to dangerous places' in J. Flint (ed.) *Housing, Urban Governance and Anti-social Behaviour*. Bristol: Policy Press.

Carr, H. and Cowan, D. (2006) 'Labelling: Constructing definitions of anti-social behaviour?' in J. Flint (ed.) *Housing, Urban Governance and Anti-social Behaviour*. Bristol: Policy Press.

Casey, R. and Flint, J. (2008) Governing through localism, contract and community: Evidence from anti-social behaviour strategies in Scotland' in *ASBO Nation: The Criminalisation of Nuisance*. Bristol: Policy Press.

Cavadino, M. and Dignan, J. (2002) *The Penal System: An Introduction*, 3rd edition. London: Sage.

Cohen, S. (1972) *Folk Devils & Moral Panic*. London: MacGibbon & Kee.

Cohen, S. (1985) *Visions of Social Control*. Cambridge: Polity Press.

Collins, E. G. C. and Blodgeth, T. B. (1981) 'Sexual Harassment…some see it…some won't', *Harvard Business Review*, March/April 59: 76–95.

Collins, S. and Cattermole, R. (2006) *Anti-Social Behaviour: Powers and Remedies*. Sweet & Maxwell: London.

Conservative Party (1979) *The Conservative Party Manifesto 1979*. London.

Cornwell, B. and Linders, A. (2002) 'The myth of "moral panic": An alternative account of LSD prohibition', *Deviant Behaviour*, 23(4): 307–30.

Cowan, D., Pantazis, C. and Gilroy, R. (2001) 'Social housing as crime control: An examination of the role of housing management in policing sex offenders', *Social and Legal Studies*, 10: 435–57.

Critcher, C. (2008) 'Moral panic analysis: Past, present and future', *Sociology Compass*, 2(4): 1127–44.

Culpitt, I. (1999) *Social Policy & Risk*. London: Sage.

de Young, Mary (2004) *The Day Care Ritual Abuse Moral Panic*. Jefferson, NC: McFarland.

Deacon, A. J. (2004) 'Justifying conditionality: The case of anti-social tenants', *Housing Studies*, 19(6): 217–34.

Dennis, N. (1993) *Rising Crime and the Dismembered Family: How Conformist Intellectuals Have Campaigned Against Common Sense*. London: Civitas.

Dennis, N. (2004) *Public Concern About Crime*. London: Civitas. Available at http://www.civitas.org.uk/pubs/concernAboutCrime.php [accessed 15 March 2006].

Dennis, N. and Erdos, G. (2005) *Cultures and Crimes: Policing in Four Nations*. London: Civitas.

Donoghue, J. (2005) *Critical Reflections on the Use of ASBOs in Scotland*. University of Glasgow, Urban Studies website, http://www.gla.ac.uk/media/media_7500_en.pdf [accessed 23 August 2006].

Donoghue, J. (2007) 'The judiciary as a primary definer on antisocial behaviour orders', *The Howard Journal of Criminal Justice*, 46(4): 417–30.

Donoghue, J. (2008) 'Antisocial behaviour orders in Britain: Contextualising risk and reflexive modernisation', *Sociology*, 42(2): 337–55.

Downes, D. and Morgan, R. (2007) 'No turning back: The politics of law and order into the new millenium', in M. Maguire, R. Morgan and R. Reiner (eds) *The Oxford Handbook of Criminology*. Oxford: Oxford University Press.

Drakeford, M., and McCarthy, K. (2000) 'Parents, responsibility and the new youth justice', in B. Goldson (ed.) *The New Youth Justice*. Russell House: Lyme Regis.

Edwards, A. and Hughes, G. (2005) 'Comparing the governance of safety in Europe: A geo-historical approach', *Theoretical Criminology*, 9(3): 259–63.

Ekberg, M. (2007) 'The parameters of the risk society: A review and exploration', *Current Sociology*, 55(3): 343–66.

Etzioni, A. (1993) *The Spirit of Community. The Reinvention of American Society*. New York: Touchstone.

Ewald, F. (1991) 'Insurance and risk', in G. Burchell, C. Gordon and P. Miller (eds) *The Foucault Effect: Studies in Governmentality*. Hemel Hempstead: Harvester Wheatsheaf.

Farley, L. (1980) *Sexual Shakedown: The Sexual Harassment of Women on the Job*. New York: Warner Books.

Ferguson, H. (1997) 'Protecting children in new times: Child protection and the risk society', *Child & Family Social Work*, 2: 221–34.

Ferguson, H. (2001) 'Social work, individualization and life politics', *British Journal of Social Work*, 31: 41–55.

Flint, J. (ed.) (2006) *Housing, Urban Governance and Anti-social Behaviour*. Bristol: Policy Press.

Forrest, R. and Kearns, A. (2001) 'Social cohesion, social capital and the neighbourhood', *Urban Studies*, 38: 2125–43.

Foucault, M. (1977) *Discipline and Punish*. London: Penguin.

Foucault, M. (2003) *Society Must Be Defended*. London: Picador.

Galligan, D. J. (1996a) *Due Process and Fair Procedures*. Oxford: Clarendon Press.

Garland, D. (2001) *The Culture of Control: Crime and Social Order in Contemporary Society*. Oxford: Oxford University Press.

Genn, H. and Genn, Y. (1990) *The Effectiveness of Representation before Tribunals*. London: Lloyd's of London Press.

Giddens, A. (1990) *The Consequences of Modernity*. Cambridge: Polity Press.

Giddens, A. (1991) *Modernity and Self-Identity*. Cambridge: Polity Press.

Giddens, A. (1992) *The Transformation of Intimacy: Sexuality, Love and Eroticism in Modern Societies*. Cambridge: Polity Press.

Giddens, A. (1994) *Beyond Left and Right: The Future of Radical Politics*. Stanford, CA: Stanford University Press.

Giddens, A. (1998) *The Third Way: The Renewal of Social Democracy*. Cambridge: Polity Press.

Gil-Robles, A. (2005) *Report by the Commissioner for Human Rights on his Visit to the UK*. Strasbourg: Council of Europe.

Gillies, V. (2005) 'Meeting parents' needs? Discourses of "support" and "inclusion" in family policy', *Critical Social Policy*, 25(1): 70–90.

Goldsmith, C. (2008) 'Cameras, cops and contracts: What anti-social behaviour management feels like to young people' in *ASBO Nation: The Criminalisation of Nuisance*. Bristol: Policy Press.

Gordon, D. and Pantazis, C. (eds) (1997) *Breadline Britain in the 1990s*. Aldershot: Ashgate.

Green, S. (2008) Rationing criminal procedure: A comment on Ashworth and Zedner, *Criminal Law and Philosophy*, 2: 53–8.

Green, D. G., Grove, E. and Martin, N. A. (2005) *Crime and Civil Society: Can We Become a More Law-Abiding People?* London: Ciritas.

Griffin, C. (1993) *Representations of Youth: The Study of Youth in Britain and America*. Cambridge: Polity

Gummesson, E. (1991) *Qualitative Methods in Management Research*. Thousand Oaks, CA: Sage.

Hall, S., Critcher, C., Jefferson, T., Clarke, J. and Roberts, B. (1978) *Policing the Crisis*. London: Macmillan.

Halliday, S. (2004) *Judicial Review and Compliance with Administrative Law*. Oxford: Hart.

Harrison, M. (2001) *Housing, Social Policy and Difference*. Bristol: Policy Press.

Haworth A. and Manzi, T. (1999) 'Managing the "underclass": Interpreting the moral discourse of housing management', *Urban Studies*, 36: 153–65.

Hayward, K. J. and Yar, M. (2006) 'The "Chav" phenomenon: Consumption, media and the construction of a new underclass', *Crime, Media, Culture*, 2(1): 9–28.

Hester, R. (2000) 'Community Safety and the New Youth Justice', in B. Goldson (ed.) *The New Youth Justice*. Lyme Regis: Russell House.

Hills, J., Le Grand, J. and Piachaud, D. (eds) (2002) *Understanding Social Exclusion*. Oxford: Oxford University Press.

Holt, A. (2008) 'Room for resistance? Parenting orders, disciplinary power and the production of the "bad parent"' in *ASBO Nation: The Criminalisation of Nuisance*. Bristol: Policy Press.

Home Office (1992) *British Crime Survey, 1992*. Research and Planning Unit, Social and Community Planning Research. London: Home Office.

Home Office (1998) *The Crime & Disorder Act 1998*. London: Home Office.

Home Office (2000a) *Antisocial Behaviour Orders: Guidance on Drawing up Local ASBO Protocol*. London: Home Office.

Home Office (2000b) *National Strategy for Neighbourhood Renewal, Report of Policy Action Team 8: Antisocial Behaviour.* London: Home Office.

Home Office (2003a) *Respect and Responsibility: Taking a Stand Against Antisocial Behaviour.* London: Home Office.

Home Office (2003b) *The One Day Count of Antisocial Behaviour.* London: Home Office.

Home Office (2004a) *The British Crime Survey 2003/04.* London: Home Office.

Home Office (2004b) *Defining and Measuring Antisocial Behaviour.* London: Home Office.

Home Office (2004c) *Together – Cost Survey of ASBOs.* London: Home Office.

Home Office (2005b) *Home Office Press Release: New Measures Will Tackle Causes of Antisocial Behaviour.* London: Home Office.

Home Office (2006a) Strengthening powers to tackle antisocial behaviour. *Consultation Paper.* London: Home Office.

Home Office (2006b) *Crime in England and Wales 2005/2006.* London: Home Office.

Home Office (2006c) Department for Constitutional Affairs and the Attorney General's Office, *Delivering Simple, Speedy, Summary Justice*, available at www.dca.gov.uk/publications/reports_reviews/delivery-simple-speedy.pdf [accessed 3 May 2007].

Home Office (2007) *Home Office Statistics on ASBOs* available at www.crimereduction.gov.uk [accessed 2 March 2007].

Home Office (2008) Home Office Statistics on ASBOs, available at www.crimereduction.gov.uk [accessed 7 October 2008].

Home Office (2009) *Home Secretary's Speech on Crime and Communities*, Home Office Website, available at http://press.homeoffice.gov.uk/Speeches/home-sec-speech-crime-09 [accessed 14 August 2009].

Hood, R. (1992) *Race and Sentencing.* Oxford: Oxford University Press.

House of Commons (1998) Home Affairs Committee Publications: Third Report, Session 1997–8, available at http://www.parliament.the-stationery-office.co.uk/pa/cm199798/cmselect/cmhaff/486/486ap01.htm [accessed 20 September 2009].

House of Commons (2005) *Home Affairs Committee – Anti social Behaviour: Fifth Report of Session 2004–05*, vol 1. London: Stationary Office.

House of Commons (2004a) *Written Evidence to the Home Affairs Committee.* London: Stationary Office.

House of Commons (2004b) *Oral Evidence to the Home Affairs Committee.* London: Stationary Office.

House of Commons (2005a) *Home Affairs Committee – Antisocial Behaviour: Fifth Report of Session 2004–05, Volume 1.* London: Stationary Office.

House of Commons (2005b) *Written Answers, 5 April 2005*, available at http://www.publications.parliament.uk/pa/cm200405/cmhansrd/vo050405/index/50405-x.htm [accessed 23 November 2005].

House of Commons (2006) *Hansard Written Answers for 5 June 2006 (pt 0640)*, available at http://www.publications.parliament.uk/pa/cm200506/cmhansrd/cm060605/text/60605w0645.htm [accessed 17 August 2006].

Housing Corporation (1998) *Safe as Houses – A Community Safety Guide for Registered Social Landlords.* London: Crime Concern.

Hunter, C., Nixon, J. and Parr, S. (2004) *What Works for Victims and Witnesses of Antisocial Behaviour*. London: Home Office.

Innes, M. (2004) 'Signal crimes and signal disorders: Notes on deviance as communicative action', *British Journal of Sociology*, 55(3): 335–55.

Innes, M. and Jones, V. (2006) *Neighbourhood Security and Urban Change*. New York: Joseph Rowntree Foundation.

Ipsos Mori (2005) *Public Concern About ASB And Support For ASBOs*. London: Ipsos Mori. Available at http://www.ipsos-mori.com/content/public-concern-about-asb-and-support-for-asbos.ashx [accessed 10 June 2006].

Jacobson, J., Millie, A. and Hough, M. (2008) 'Why tackle anti-social behaviour?' in *ASBO Nation: The Criminalisation of Nuisance*. Bristol: Policy Press.

Jewkes, Yvonne (2004) *Media and Crime: Key Approaches to Criminology*. London, UK: Sage.

Jones, T. and Newburn, T. (2007) *Policy Transfer and Criminal Justice: Exploring US Influence over British Crime Control Policy*. Maidenhead: Open University Press.

Judicial Studies' Board (2007) *ASBOs: Guidance for the Judiciary*, 3rd edn. London: JSB.

Justices' Clerks' Society (JCS) (2006) *Antisocial Behaviour Orders: Good Practice Guide*. London: JCS.

Kania, R. R. E. (1988) 'Conservative ideology in criminology and criminal justice', *American Journal of Criminal Justice*, 8(1): 74–96.

Karmen, A. (2001) *New York Murder Mystery: The True Story Behind the Crime Crash of the 1990s*. New York: New York Press.

Kelling, G. L. and Sousa, W. H. Jr (2001) *Police Matter? An Analysis of the Impact of New York City's Police Reforms*. New York: Manhattan Institute.

Kemshall, H. and MacGuire, M. (2001) 'Public protection, partnership and risk penalty', *Punishment and Society*, 3(2): 163–90.

Kemshall, H. (1997) 'Concepts of risk in relation to organizational structure and functioning within the personal social services & probation', *Social Policy and Administration*, 31(3): 213–32.

Kinloch, G. C. (1981) *Ideology and Contemporary Sociological Theory*. Englewood Cliffs, NJ: Prentice-Hall.

Kors, A. C. and Silvergate, H. A. (1998) *The Shadow University: The Betrayal of Liberty on America's Campuses*. New York: The Free Press.

Kramer, R. C. (1982) *Teaching Critical Criminology to Criminal Justice Students: A Dilemma and a Proposed Resolution*. Paper presented to the annual meeting of the American Society of Criminology, Toronto, Canada, November 1982.

Labour Party (1974) *The Labour Party Manifesto 1974*. London: Labour Party.

Lash, S. and Urry, J. (1994) *Economics of Signs & Spaces*. London: Sage.

Lee, R. (2000) *Unabtrusive Methods in Social Research*. Buckingham: Open University Press.

Lianos, M. (2003) 'Social control After Foucault', *Surveillance and Society*, 1(3): 412–30.

Liberty (2004) *Liberty's evidence to the Home Affairs Committee on Antisocial Behaviour*. London: Liberty.

Lister, D. (2006) 'Tenancy agreements: A mechanism for governing anti-social behaviour?' in *Housing, Urban Governance and Anti-Social Behaviour*. Bristol: Policy Press.

Loader, I. and Walker, N. (2007) *Civilizing Security*. Cambridge: Cambridge University Press.

Lyotard, J. (1979) *The Postmodern Condition: A Report on Knowledge*. Manchester: Manchester University Press.

Macdonald, S. (2003) 'A suicidal woman, roaming pigs and a noisy trampolinist: Refining the ASBO's definition of 'antisocial behaviour', *The Modern Law Review*, 69(2): 183–213.

Macdonald, S. (2006) 'The principle of composite sentencing: Its centrality to, and implications for, the ASBO', *Criminal Law Review*, 791–808.

MacKinnon, C. A. (1979) *Sexual Harassment of Working Women: A Case of Sex Discrimination*. New Haven, CT: Yale University Press.

Madge, N. (2004) 'Anti-social behaviour orders: Case law reviewed', *Legal Action*, December: 20–2.

Maruna, S. and King, A. (2004) 'Public opinion and community penalties', in A. Bottoms, S. Rex and G. Robinson (eds), *Alternatives to Prison: Options for an Insecure Society*. Collumpton: Willan.

Mason, D. (2005) 'ASBOs – use and abuse', *New Law Journal*, 155 (7161) January: 129.

Mayfield, G. and Mills, A. (2008) 'Towards a balanced and practical approach to anti-social behaviour management' in *ASBO Nation: The Criminalisation of Nuisance*. Bristol: Policy Press.

McInnes, J. (2005) (unpublished) *Summary Criminal Justice: How We Got Here – Seeing the Bigger Picture*. Paper presented to the Antisocial Behaviour and Summary Justice Reform Conference. Edinburgh, 21 November 2005.

McIntosh, B. (2008) 'ASBO youth: Rhetoric and realities', in *ASBO Nation: The Criminalisation of Nuisance*. Bristol: Policy Press.

Miller, J. M. (1996) 'Scholar to scholar: Evaluating criminology and criminal justice with Richard A. Wright', *American Journal of Criminal Justice*, 21(1): 105–16.

Miller, D. and Kitzinger, J. (1998) 'AIDS, the policy process and moral panics', in D. Miller, J. Kitzinger, K. Williams and P. Beharrell (ed.) *The Circuit of Mass Communication: Media Strategies, Representation and Audience Reception in the AIDS Crisis*. London: Sage, 213–22.

Millie, A., Jacobson, J., McDonald, E. and Hough, M. (2005) *Anti-social Behaviour Strategies: Finding a Balance*. Oxon: The Policy Press.

Morgan, R. (1997) 'Imprisonment', in M. Maguire, R. Morgan and R. Reiner (eds), *The Oxford Handbook of Criminology*. Oxford: Oxford University Press.

Morgan, R. and Newburn, T. (2007) 'Youth Justice', in M. Maguire, R. Morgan and R. Reiner (eds) *The Oxford Handbook of Criminology*. Oxford: Oxford University Press.

Morrison, R. (2008) 'For once, let's hear it for the boys' in *The Times*, published 14 May 2008, available at http://www.timesonline.co.uk/tol/comment/article3923999.ece [accessed 14 May 2008].

National Association of Probation Officers (NAPO) (2004) *Antisocial Behaviour Orders: Analysis of the First Six Years*. London: NAPO.

National Audit Office (NAO) (2006) *The Home Office: Tackling Antisocial Behaviour*. London: NAO.

National Housing Federation (NHF) (1999) *Housing and Crime: Safe as Houses*. London: NHF.

Nixon, J., Hunter, C. and Shayer, S. (1999) *The Use of Legal Remedies by Social Landlords to Deal with Neighbourhood Nuisance*, Centre for Regional Economic & Social Research: Sheffield Hallam University.

Nixon, J. and Parr, S. (2006) 'Anti-social behaviour: Voices from the front line', in *Housing, Urban Governance and Anti-Social Behaviour*. Bristol: Policy Press.

Ohlin, L. D. (1971) 'The effect of social change on crime and law enforcement', in L. W. Levy (ed.) *Perspectives on the Report of the President's Commission on Law Enforcement and the Administration of Justice*. New York: Da Capo Press.

Parr, S. (2006) *Intensive welfare support: A new approach to the most challenging families?* Paper presented at ENHR Conference, Ljubljana, July.

Parton, N. (1998) 'Risk, advanced liberalism and child welfare', *British Journal of Social Work*, 28(1): 5–27.

Pawson, H. and McKenzie, C. (2006) 'Social landlords, anti-social behaviour and countermeasures', in *Housing, Urban Governance and Anti-Social Behaviour*. Bristol: Policy Press.

Pearson, G. (1983) *Hooligan: A History of Respectable Fears*. London: Macmillan.

Posner, R. (1992) *Economic Analysis of Law* (3rd edn). Boston, MA: Little Brown.

Pound, R. (2002 [1931]) *The Call for Realist Jurisprudence*. Oxford: Oxford University Press.

Power, M. (1997) *The Audit Society: Rituals of Verification*. Oxford: Oxford University Press.

Power, A. and Mumford, K. (1999) *The Slow Death of Great Cities? Urban Abandonment or Urban Renaissance?* London: The Joseph Rowntree Foundation. Reproduced by permission of the Joseph Rowntree Foundation.

Pratt, J., Brown, D., Brown, M., Hallsworth, S. and Morrison, W. (eds) (2005) *The New Punitiveness: Trends, Theories, Perspectives*. Cullompton: Willan.

Rawlings, P. (1992) *Drunks, Whores and Idle Apprentices: Criminal Biographies of the Eighteenth Century*. London: Routledge.

Raynor, P. (2007) 'Community Penalities: Probation, "What Works" and offender management', in M. Maguire, R. Morgan and R. Reiner (eds) *The Oxford Handbook of Criminology*, fourth edition, 1061–99.

Reiner, R. (2000) 'Crime and control in Britain', *Sociology*, 34: 71–94.

Ren, L., Zhao, J. and Lovrich, N. P. (2008) 'Liberal versus conservative public policies on crime: What was the comparative track record during the 1990s?' *Journal of Criminal Justice*, 4: 316–25.

Roberts, J. (2006) *Crime and Justice in Britain: Trends, Analysis and Conclusions*. Available at http://www.number10.gov.uk/Page9705 [accessed 22 August 2001]. London: Prime Minister's Office.

Rosch, J. (1985) 'Crime as an Issue in American Politics', in E. S. Fairchild and V. J. Webb (eds) *The Politics of Crime and Criminal Justice*. Beverly Hills, CA: Sage.

Rose, N. (1999) *Powers of Freedom: Reframing Political Thought*. Cambridge: Cambridge University Press.

Rowe, M. (1981) 'Dealing with sexual harassment', *Harvard Business Review*, 59(3): 42–7.

Sampson, R. J. and Raudenbush, S. W. (2004) 'Seeing disorder: Neighbourhood stigma and the social construction of "Broken Windows"', *Social Psychology Quarterly*, (67)4: 319–42.

SANE (2005) 'Mental health charities slam ASBO case', *Community Care*, 28 February.

Saugeres, L. (2000) 'Of tidy gardens and clean houses: Housing officers as agents of social control', *Geoforum*, 13: 587–99.

Scheingold, S. A. (1984) *The Politics of Law and Order.* New York: Longman.

Schillewaert, N., Langerak, F. and Duhamel, T. (1998) 'Non probability sampling for www surveys: A comparison of methods', *Journal of the Market Research Society*, 4(40): 307–13.

Scott, S. and Parkey, H. (1998) 'Myths and reality: Neighbour nuisance problems in Scotland', *Housing Studies*, 13(3): 325–45.

Scottish Executive (2001) *The Scottish Crime Survey 2000.* Edinburgh: Scottish Executive.

Scottish Executive (2002) *The Scottish Household Survey 2001/02.* Edinburgh: Scottish Executive.

Scottish Executive (2003) *Putting our Communities First: A Strategy for Tackling Antisocial Behaviour.* Edinburgh: Scottish Executive.

Scottish Executive (2004a) *Antisocial Behaviour Etc. (Scotland) Act 2004: Guidance on Antisocial Behaviour Strategies.* Edinburgh: Scottish Executive.

Scottish Executive (2004b) *Antisocial Behaviour Etc. (Scot) Act 2004.* Edinburgh: Scottish Executive.

Scottish Executive (2005a) The Use of Antisocial Behaviour Orders in Scotland (Part 1). Edinburgh: Scottish Executive.

Scottish Executive (2005b) *The Use of Antisocial Behaviour Orders in Scotland (Part 2).* Edinburgh: Scottish Executive.

Scottish Office (1999) (unpublished) *The Cost of Antisocial Behaviour.* Edinburgh: Scottish Office.

Scottish Parliament (2004) *Report on the Financial Memorandum of the Antisocial Behaviour Etc. (Scotland) Bill.* Available at http://www.scottish.parliament.uk/business/committees/finance/reports-04/fir04-finmemo-01.htm [accessed on 25 March 2006].

Scottish Parliamentary Written Answer, 2 March 2007. Available at http://www.scottish.parliament.uk/business/pqa/wa-07/wa0302.htm [accessed on 29 April 2007].

Scourfield, J. and Welsh, I. (2003) 'Risk, reflexivity and social control in child protection: New times or same old story?' *Critical Social Policy*, 23(3): 398–420.

Scraton, P. (2004) 'Streets of terror: Marginalization, Criminilization, and Authoritarian Renewal', Social Justice, 31(2): 130–58.

Sennett, R. (1996) *The Uses of Disorder: Personal Identity and City Life.* London: Faber and Faber.

Sennett, R. (2003) *Respect: The Formation of Character in an Age of Inequality.* London: Penguin.

Sentencing Guidelines Council (2007) Guidelines on Breach of an ASBO, available at http://www.sentencing-guidelines.gov.uk/docs/Advice%20%20Breach%20of%20an -%20Anti-Social%20Behaviour%20Order.pdf [accessed 17 September 2008].

Shaw, C. (1931) *The Natural History of the Delinquent Career.* Chicago, IL: University of Chicago Press.

Simester, A. P. and von Hirsch, A. (2006) *Incivilities: Regulating Offensive Behaviour.* Oxford: Hart.

Simon, J. (2008) 'Review of policy transfer and criminal justice: Exploring US influence over British crime control policy', in Trevor Jones and Tim Newburn (eds), Maidenhead: Open University Press, 2007, *British Journal of Criminology*, 48: 251–4.

Social Exclusion Unit (SEU) (2000a) *National Strategy for Neighbourhood Renewal – Report of Policy Action Team 8: Antisocial Behaviour*, accessible at http://socialexclusionunit.gov.uk/publications/pat/pat8/ondex.htm [accessed 3 May 2005].

Social Exclusion Unit (SEU) (2000b) *National Strategy for Neighbourhood Renewal – Report of Policy Action Team 13: Shops*, available at http://www.doh.gov.uk [accessed 3 May 2005].

Sommerlad, H. (2004) 'Some reflections on the relationship between citizenship, access to justice, and the reform of legal aid', *Journal of Law and Society*, 31(3): 345–68.

Spelman, W. (1994) *Criminal Incapacitation*. New York: Plenum.

Squires, P. and Stephen, D. (2005) *Rougher Justice: Antisocial behaviour and Young People*. Collumpton: Willan.

Squires, P. (2006) 'New Labour and the Politics of Antisocial Behaviour', *Critical Social Policy*, 26(1): 144–68.

Squires, P. (ed.) (2008) *ASBO Nation: The Criminalisation of Nuisance*. Bristol: Policy Press.

Stenson, K. and Edwards, A. (2004) 'Policy transfer in local crime control: Beyond naïve emulation', in T. Newburn and R. Sparks (eds) *Criminal Justice and Political Cultures: National and International Dimensions of Crime Control*. Collumpton: Willan.

Stephen, D. (2008) 'The responsibility of respecting justice: An open challenge to Tony Blair's successors' in *ASBO Nation: The Criminalisation of Nuisance*. Bristol: Policy Press.

Steventon, G. (2006) *Complicity or Coercion? Social Control and the Antisocial Behaviour Order*. Paper presented at the BCS Conference, July.

Swoboda, S. J., Muehlberger, N., Weitkunat, R. and Schneeweiss, S. (1997) 'Web-based surveys by direct mailing: An innovative way of collecting data', *Social Science Computer Review*, 15(3): 242–55.

The Guardian, Blair's Respect Agenda Ditched Claim Tories, 24 December 2007.

The Independent, Crime in Britain: Rural areas bear brunt of criminal offensive: Overall rate of increase slows – 'Panic' reforms attacked – Fewer murders confirm trend, 29 April 1993.

The Times, Alan Johnson to tackle antisocial behaviour as top priority, 22 June 2009.

Tonry, M. (2004) *Punishment and Politics: Evidence and Emulation in the Making of English Crime Control Policy*. Cullompton: Willan.

Upson, A. (2006) *Perceptions and Experiences of Antisocial Behaviour: Findings from the 2004/05 British Crime Survey, Home Office Report*. London: Home Office.

Vaughn, B. (2000) 'The Government of Youth: Disorder & Dependence', *Social & Legal Studies*, 9(3): 347.

Wain, N. (2007) *The ASBO: Wrong Turning, Dead End*. London: Howard League for Penal Reform.

Waiton, S. (2008) 'Asocial not anti-social: The "Respect Agenda" and the "therapeutic me"' in *ASBO Nation: The Criminalisation of Nuisance*. Bristol: Policy Press.

Walker, S. (1985) *Sense and Nonsense About Crime: A Policy Guide*. Monterey, CA: Brooks/Cole.

Walker, S. (1998) *Sense and Nonsense about Crime: A Policy Guide* (4th edn). Belmont, CA: Wadsworth.

Walker, A. L., Lind, A. and Thibaut, J. (1979) 'The Relation between procedural and distributive justice', *Virginia Law Review*, 65: 1401–20.

Walters, R. and Woodward, R. (2007) 'Punishing "poor parents": "respect", "responsibility" and Parenting Orders in Scotland', *Youth Justice*, 7(1): 5–20.

Webb, S. (2006) *Social Work in a Risk Society: Social and Political Perspectives*. Basingstoke: Palgrave Macmillan.

White, S. (1999) 'Rights and responsibilities: A social democratic perspective', in A. Gamble and T. Wright (eds) *The New Social Democracy*. Oxford: Blackwell.

Whitehead, M. (2004) 'The urban neighbourhood and the moral geographies of British urban policy', in Johnstone, C. and Whitehead, M. (eds) *New Horizons in British Urban Policy*. London: Ashgate.

Williams, R. J. (2004) *The Anxious City*. London: Routledge.

Wilson, J. Q. (1975) *Thinking About Crime*. New York: Random House.

Wilson, D. and Ashton, J. (1998) *What Everyone in Britain Should Know About Crime and Punishment*. London: Blackstone Press.

Wilson, J. and Kelling, G. (1982) 'Broken windows: The police and neighbourhood safety', *The Atlantic Monthly*, March: 29–37.

Young, J. (1971) *The Drugtakers: the Social Meaning of Drug Use*. London: McGibbon and Kee.

Young, J. (2001) 'Identity, community and social exclusion', in R. Matthews and J. Pitts (eds), *Crime, Disorder and Community Safety*. London: Routledge.

Youth Justice Board (2006) *Antisocial Behaviour Order Research*. London: YJB.

Zedner, L. (2002) 'Dangers of dystopias in penal theory', *Oxford Journal of Legal Studies*, 22(2): 341–66.

Zedner, L. (2005) 'Securing liberty in the face of terror: Reflections from criminal justice', *Journal of Law and Society* 32(4): 507–33.

Legislation

Antisocial Behaviour Act 2003
Antisocial Behaviour Etc. Act (Scotland) 2004
Crime and Disorder Act 1998
Criminal Justice (Scotland) Act 2003
Criminal Procedure (Scotland) Act 1995
European Convention on Human Rights
Human Rights Act 1998
Magistrates' Courts Act 1980
NHS and Community Care Act 1990
Police Reform Act 2002
Protection from Harassment Act 1997
Serious Organised Crime and Police Act 2005
Sex Offenders Act 2003
Scotland Act 1998
Supreme Court Act 1981

Cases

Aberdeen City Council v Fergus (2006) GWD 36

Clingham v Kensington and Chelsea RLBC [2003] HLR 17

Coventry City Council v Finnie [1995] QBD 432

Edinburgh City Council v Donald Gibson (2006)

Gillbard v Cardon District Council [2006] EWHC 3233 (Admin) (cited in JSB, 2007: 17)

Glasgow Housing Association Ltd v O'Donnell (2004) GWD 29

Glasgow Housing Association Ltd v Sharkey (2004)

Hills v Chief Constable of Essex Police [2006] EWHC 2633 (Admin)

Hussain v Lancaster City Council [2000] QBD 1

Kenny v Leeds Magistrates' Court [2004] EWCA Civ 312

Manchester City Council v Lee [2004] HLR 11 161

Manchester City Council v Romano [2004] EWCA Civ 834

Malzy v Eicholz [1916] 2 KB 308

Northampton BC v Lovat [1997] 96 LGR

O'Reilly v Mackman [1983] 1 AC 237

R (A) v Leeds' Magistrates Court and Leeds City Council [2004] EWHC 554 (Admin)

R v Adam Lawson [2006] 1 Cr App R (S) 323

R v Boness [2005] EWCA Crim 2395

R v H, Stevens and Lovegrove [2006] EWCA Crim 255

R (Keating) v Knowsley Metropolitan Borough Council [2004] EWHC 1933 (Admin)

R v Kirby [2005] EWCA Crim 1228

R (Manchester City Council) v Manchester City Magistrates' Court [2005] EWCH 253 (Admin)

R (McCann) v Manchester Crown Court [2003] 1 AC 787

R v Morrison [2006] 1 Cr. App. R (s) 488 (85)

R v P [2004] EWCA Crim 287

R v Tripp [2005] EWCA Crim 2253

R (W) v Acton Youth Court [2005] EWCH 954 (Admin)

R v W and F [2006] EWCA Crim 686

R v Williams [2006] 1 Cr App R (S) 305

Re T (An Adult) (Consent to Medical Treatment) [1992] 2 FLR 458

Smith v Scott [1973] Ch. 314

Stanley, Marshall and Kelly v The Commissioner of Police for the Metropolis and The Chief Executive of Brent Council [2004]

W v DPP [2005] EWHC Admin 1333

Index

Note: Locators in italics followed by *n* indicate notes, locators in italics followed by *f* indicate figures and locators in italics followed by *t* indicate tables.